Ingenieurwissenschaftliche Bibliothek
Engineering Science Library

Herausgeber / Editor: István Szabó, Berlin

I. B. Gertsbakh · Kh. B. Kordonskiy

Models of Failure

Translation from the Russian

Springer-Verlag New York Inc. 1969

Professor I. B. GERTSBAKH · Professor DR. KH. B. KORDONSKIY

Institute of Civil Aviation Engineers of Riga / USSR

Title of the Russian Original Edition:

Modeli Otkazov

Biblioteka Inzhenera po Nadezhnosti

Edited by B. V. Gnedenko

Publisher: Sovetskoe Radio · Moscow 1966

Translated by Scripta Technica Inc., Washington, USA

With 68 Figures and 3 Tables

ISBN 978-3-540-04569-4   ISBN 978-3-642-87519-9 (eBook)
DOI 10.1007/978-3-642-87519-9
Library of Congress Catalog Card Number 73-86177
Title-No. 4250

Preface

The increase in the requirements on the reliability of units makes
it necessary to analyze the relationship between mathematical methods of calculating reliability and the physical nature of failures. The difficulty of such an analysis is obvious. On the one
hand, in making a representation of the physical picture of a phenomenon, one can make an error in the direction of excessive simplification. On the other hand, in the mathematical treatment of
the physical scheme, it may be necessary to use extremely complex
and fine analytical methods, and their simplified exposition bor-
ders on vulgarization.

Without the aid of a large number of specialists working in the
field of analysis and calculation of systems reliability, an exposition of models of failures and their mathematical treatment
would be unobtainable.

The authors take this opportunity to express their gratitude to
Academicians N. G. B r u y e v i c h and Y u. V. L i n n i k ,
conversations with whom clarified a number of problems treated, to
active member of the Academy of Sciences of the UkrSSR B. V.
G n e d e n k o and Professor Y a. B. S h o r , without whose
support the work on the book would have been impossible, to acting
member of the Academy of Sciences of the UkrSSR S. V. S e r e n -
s e n , who was of help in the treatment of the complicated problems of accumulation of aging of metals, and to Professors I. A.
I b r a g i m o v , A. M. K a g a n , and A. D. S o l o v ' -
y e v , who made a number of valuable suggestions.

Contents

Introduction

For each unit, the properties that it must possess in the course of its use are listed. A deviation in the properties of units from the prescribed conditions is considered as a fault. A state of fault is denoted by the term "failure". Failures may differ one from another as regards their significance in use. Therefore, in what follows, when we use the word "failure", we shall mean faults of approximately the same significance in use.

Among failures, we distinguish between removable failures, that is, those that can be eliminated when the unit is subjected to a renewal by means of maintenance, and nonremovable ones, that is, those such that the unit must be taken out of use. The time of operation between two removable failures is called the lifetime. In what follows, we shall denote this time by τ. For simplicity, we shall assume that the repairs possess the property of equivalence, that is, that a repair can be regarded as replacement of a unit that has failed with an equivalent new one.

The lifetime τ of an arbitrary unit has random variations. Therefore, the calculations and tests of reliability are connected with the use of the methods of probability theory and mathematical statistics. We assume that the reader is familiar with the fundamentals of these methods. To make the text easier to read, we shall give below the notations and concepts used in it.

We denote by τ_i the observed lifetime of a unit. Consequently,

$$\overline{\tau} = \frac{1}{N}\left(\tau_1 + \tau_2 + \ldots + \tau_N\right) \tag{1}$$

is the mean lifetime, calculated from the results of observation of N exemplaires of the given unit.

We denote by F(T) the distribution function of the time τ:

$$F(T) = \underline{P}\{\tau \leq T\} \qquad (2)$$

where $\underline{P}\{c\}$ denotes the probability of the event c.

We denote by m(T) the number of units that fail in a period of time T. Therefore, the frequency of the event $\tau \leq T$ is represented by

$$\nu(T) = \frac{m(T)}{N}, \qquad (3)$$

where N is the number of units observed.

In estimating the parameters of the theoretical distribution from experimental data, we shall use extensively the equation

$$F(T) = \nu(T), \qquad (4)$$

which is of a statistical nature and is the more exact the greater the number N of units observed.

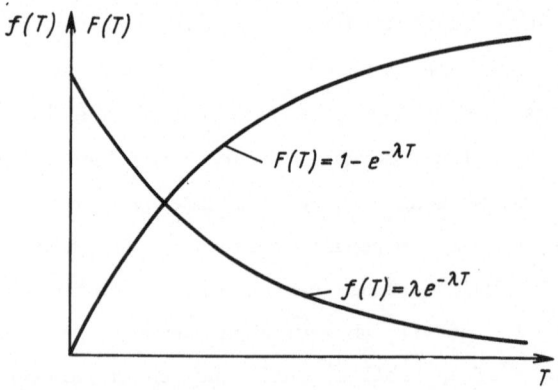

Fig. 1. The distribution function and the probability density of an exponential distribution.

The probability distribution density of the quantity τ is denoted by f(T). If the function F(T) has a derivative, then

$$f(T) = \frac{dF(T)}{dT}. \qquad (5)$$

Fig. 1 shows the graphs of the function F(T) and its density f(T) for one of the more frequently encountered distributions of the time τ, known as the exponential distribution.

Of course,

$$\underline{P}\{\tau \leq T\} = F(T) = \int_0^T f(t)dt. \tag{6}$$

We denote by ν_k the initial k^{th}-order moment of the random variable τ;

$$\nu_k = \int_0^\infty t^k f(t)dt; \tag{7}$$

We denote its kth-order central moment by μ_k:

$$\mu_k = \int_0^\infty (t - \nu_1)^k f(t)dt. \tag{8}$$

We denote by $\underline{M}\{\tau\}$ the mathematical expectation of the time τ, which is the first initial moment. For large N, we can write the statistical equation

$$\underline{M}\{\tau\} = \overline{\tau}. \tag{9}$$

The second central moment, known as the variance, is denoted by $\underline{D}\{\tau\}$. We have

$$\underline{D}\{\tau\} = \nu_2 - \nu_1^2. \tag{10}$$

The variance of the observed values of τ_i is denoted by s_τ^2; that is,

$$s_\tau^2 = \frac{1}{N-1} \sum_{i=1}^N (\tau_i - \overline{\tau})^2. \tag{11}$$

The standard deviation of the quantity τ is denoted by $\sigma_\tau = \sqrt{\underline{D}\{\tau\}}$. When the number of observations is large, we may take

$$\sigma_\tau = s_\tau. \tag{12}$$

The experimental analogue of the kth-order initial moment is the quantity

$$n_k = \frac{1}{N} \sum_{i=1}^N \tau_i^k. \tag{13}$$

When the number N of observations is large, we can write the statistical equation

$$\nu_k = n_k. \tag{14}$$

3

In what follows, we shall have occasion to use the statistical analogue of the third-order central moment, which is calculated in accordance with the formula

$$m_3 = \frac{N}{(N-1)(N-2)} \sum_{i=1}^{N} (\tau_i - \overline{\tau})^3 , \qquad (15)$$

Here, for large values of N, we may take

$$\mu_3 = m_3 . \qquad (16)$$

Chapter I

Classifications of Causes of Failures

Failures of a unit can be classified according to various criteria. One of the more frequently used criteria is the place of failure. With this classification, we need to identify the unit as a whole, its connections, various details regarding the connections, and various elements of the details. In what follows, we shall single out only two categories of objects: a system, which is a device containing several parts, and an element, which constitutes some part of a system.

Thus, a radio receiver can be regarded as a system, whereas the amplifiers, speakers, and Vernier dial can be regarded as elements. The amplifier can, in turn, be regarded as a system the elements of which are the lamps, condensers, and so forth.

A classification of defects according to the place of failure enables us to estimate the weak point in a system and to take measures to strengthen it. It is no less important to find the reason for the occurrence of a failure at a given point. These reasons can be classified into the following groups:

Construction defects. Failures of this group arise as a consequence of an imperfection in the construction. A typical example is non-consideration of "peak" loads.

A load acting on a system and its elements usually has random variations. In the construction, one tries to keep in mind the possibility of occurrence of "peak" loads, that is, loads considerably exceeding the loads due to normal use. If an analysis and calculation of the loads are made with insufficient care, then the action of "peak" loads will lead to failures. From the point of view of analysis and calculation of reliability, it is important to have the defects in the construction show up to the same extent in all copies of the system or element under consideration.

Technological defects. Failures of this class occur as a consequence of violation of the technological manufacturing procedure chosen for the system or unit. The quality of the individual units and connections and of a unit as a whole has unavoidable random variations. Quality variations kept within sufficiently restricted limits do not show up appreciably in the reliability of the system. With sharp fluctuations in the quality, the reliability of certain items will prove considerably less than the reliability of others. Therefore, technological defects decrease the reliability of some of the items in the total set of manufactured systems or units.

Defects due to improper use. For every system, restrictions are made on the conditions of its use (restrictions on the temperature, on the frequency of vibrations, etc.) and rules are given for maintenance of the system and its parts, and so forth. Violation of the rules of use lead to premature failures; that is, they increase the speed at which the system ages. Usually, such violations affect only certain used exemplaires of the system.

Aging (wear and tear) of a system. No matter how good the quality of the unit and the system as a whole, a gradual aging (wear) is inevitable. During the course of use and storage, irreversible changes take place in metals, plastics, and other materials and the accumulative effect of these changes destroys the strength, coordination, and interaction of the parts and, in the final analysis, causes failures.

Thus, variations in the lifetime are caused by variations in the quality of the manufacture, the conditions of use, and aging processes. Here, we have the following schemes for the occurrence of failures.

The scheme of instantaneous injuries. Figure 2 shows the graph of the change in the load (stress) in one of the units of a glider in flight. The presence of individual "peaks" in the stress is typical. If we assume that a unit fails when the load S exceeds a certain level S_1, then the instant of failure is, by virtue of the

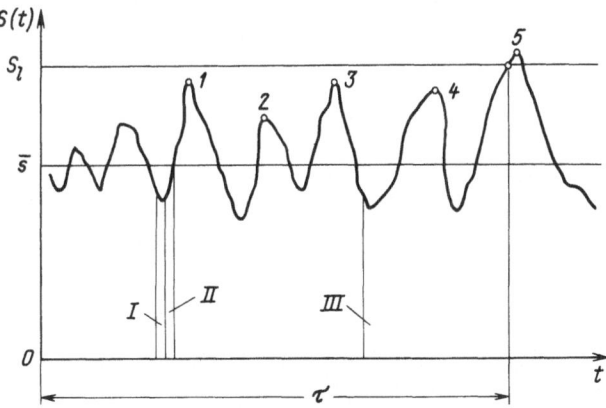

Fig. 2. A graph of the change in the stress during a flight.

randomness of the change in the load, also random. Here, it is typical that a failure in a unit occurs independently of how long it
has been in use and of what condition it is in. A striking example
of a failure of this type is a puncture in an automobile tire. It
occurs as the result of the existence of a sharp object in the path
of the tire and is independent of the condition both of the tire
and the automobile.

A scheme of accumulated changes. This scheme corresponds to the situation in which a failure occurs as the result of the gradual accumulation of injuries (gradual aging or wear). For a number of operating parameters of a unit or system, admissible limits are established in advance. A parameter's getting outside these limits is
classified as a failure. A change in the parameters is caused by
aging of a unit, and the time until a parameter gets beyond the
admissible limit is the lifetime.

A scheme of relaxation. The gradual accumulation of damage may be
not the direct but an indirect cause of a failure. Thus, the wear
of a slider and guides increases the gap between them and, under
unfavorable external conditions, this may cause them to wedge. When
there is no scheme of accumulated damage, we do not have in this
case an admissible limit for the operating parameter (the gap). The
accumulation of damage leads only to increase in the probability of
failure. In the case of this example, the combination of gradual

accumulation of damage with a sudden change in the state of the object is typical. We might also mention cases of violation of a statistically indeterminable system under the action of cyclic loads or the failure of a stand-by electronic system when the load on the operating units increases whenever the stand-by elements fail. The sudden change in the state of the object due to the gradual accumulation of damage is called relaxation.

The scheme of action of several independent causes. A situation in which several causes of failures act simultaneously is the most typical one in practice. Let us again look at the automobile tire. Obviously, at least two causes for failure occur simultaneously: a puncture due to the existence of a sharp object in its path and the gradual wear of the tire. The situation is analogous to failures in vacuum tubes. These can fail both because of a random "peak" load and because of gradual aging.

Here, it is important to note the following facts: The causes of failures often include one or two that are predominant. The influence of the other causes is so weak that they are virtually not observed in practice. Therefore, in investigating the reliability of an object, we seek first of all to ascertain the predominant causes of failures and only then do we calculate the influence of the remaining ones if this is necessary. On the other hand, if many causes of approximately equal influence act simultaneously, then the cumulative effect of these causes can be conditionally replaced with the action of a single cause that, in a certain sense, is equivalent to the entire set of causes of failure.

These schemes could be supplemented by a scheme of action of several connected (dependent) reasons for failures of a system. However, the construction of such schemes requires consideration of a large number of possible relationships between the causes, and this would make any even approximatetely complete description of them impossible. Furthermore, if the causes of failures are closely connected with each other, it makes sense to assume that only one of them is acting and to ascribe all failures to it.

8

It what follows, we shall give a mathematical description to each of these schemes. This will enable us to clarify the significance of the scheme, to show more precisely its physical meaning, and to establish for each scheme those distribution laws for the lifetime that correspond to its nature.

Chapter II

Instantaneous Damage

The exponential distribution

When an airplane lands, a radio receiver on it undergoes inertial stresses. A rough landing may cause these stresses to be considerably in excess of what is admissible and this damages or breaks units of the receiver; that is, it leads to failure of the receiver. The rough landing is a consequence either of piloting errors or difficult meteorological conditions and thus is independent of the state of the receiver. Such a failure in the receiver corresponds to the scheme of instantaneous damages, which we considered above. Let us denote by γ_i the probability of a rough landing when the airplane lands for the ith time, counted from the first flight of the airplane with the radio receiver in question on board. In accordance with our description, the receiver will get out of order with the plane's first rough landing. Let us calculate the probability that the receiver will fail on the kth landing.

For the receiver to fail on the kth landing, it is necessary and sufficient that this be the first rough landing of the plane since the radio receiver was installed in it. Let us denote by H_i the event that the ith landing is a normal one and let us denote by Γ_i the event that the ith landing is rough. Obviously, the events H_i and Γ_i are mutually exclusive and complementary; that is,

$$\underline{P}\{H_i\} + \underline{P}\{\Gamma_i\} = 1, \tag{17}$$

Furthermore, the result of the ith landing is independent of the results of the preceding landings; that is, the events H_i for different values of i are mutually independent. We denote by B_k the event that a crude landing will occur for the first time when the plane lands the kth time. Obviously, this will occur if and only if $k-1$ smooth landings have occurred but the kth landing proves rough. This is written

$$B_k = H_1 H_2 \ldots H_{k-1} \Gamma_k. \tag{18}$$

In accordance with the theorem on the multiplication of independent events,

$$\underline{P}\{B_k\} = \underline{P}\{H_1\}\underline{P}\{H_2\}\ldots\underline{P}\{H_{k-1}\}\underline{P}\{\Gamma_k\}. \tag{19}$$

Using the notation

$$\underline{P}\{\Gamma_i\} = \gamma_i, \quad \underline{P}\{H_i\} = 1 - \gamma_i, \tag{20}$$

we have

$$\underline{P}\{B_k\} = (1 - \gamma_1)(1 - \gamma_2)\ldots(1 - \gamma_{k-1})\gamma_k. \tag{21}$$

The physical meaning of our problem is such that, if the skill of the pilot remains unchanged, the probability γ_k of a rough landing remains unchanged from one landing to another. Then, for $\gamma_i = \gamma$, with $i = 1, 2, \ldots$, we have

$$\underline{P}\{B_k\} = (1 - \gamma)^{k-1}\gamma. \tag{22}$$

We denote by τ the random variable representing the number of landings that do not damage the receiver. The last formula expresses the probability that $\tau = k - 1$:

$$\underline{P}\{\tau = k - 1\} = (1 - \gamma)^{k-1}\gamma, \quad k \geq 1. \tag{23}$$

This equation gives the so-called geometric distribution of the random variable τ. The name "geometric" is connected with the fact that the right-hand member of equation (23) recalls the expression for the general term of a geometric progression.

In our present example, the random variable τ is simply the lifetime expressed as the number of landings. It is of interest to find the probability that the plane will make K landings without causing the radio receiver to fail, that is, the probability of the event $\tau > K$.

Obviously,

$$\underline{P}\{\tau > K\} + \underline{P}\{\tau \leq K\} = 1. \tag{24}$$

11

The probability $\underline{P}\{\tau \leq K\}$ is calculated in accordance with the theorem on addition of probabilities:

$$\underline{P}\{\tau \leq K\} = \sum_{n=0}^{K} \underline{P}\{\tau = n\}. \tag{25}$$

With the aid of (23), we obtain

$$\underline{P}\{\tau \leq K\} = \gamma \sum_{n=0}^{K} (1 - \gamma)^n = 1 - (1 - \gamma)^{K+1}. \tag{26}$$

Consequently, from (24),

$$\underline{P}\{\tau > K\} = (1 - \gamma)^{K+1}. \tag{27}$$

This equation enables us to calculate the probability of failure-free operation throughout an "interval of time" equal to K.

Example 1. With each landing, the probability of a rough landing is $\gamma = 0.001$. Find the probability that the plane will make 1,000 landings without damaging the radio receiver. In accordance with (27),

$$\underline{P}\{\tau > 1000\} = (1 - 0.001)^{1001}.$$

The calculations will be considerably simplified if we use the formula

$$\underline{P}\{\tau > 1000\} = (1 - \gamma)^{K+1} \approx e^{-K\gamma}. \tag{28}$$

This formula coincides very closely with the preceding one if γ is small, K great, and the product $K\gamma$ somewhere in the interval [0.1, 20]. One can show that the error incurred by replacing $(1 - \gamma)^n$ with $e^{-n\gamma}$ is of the order $n\gamma^2/2$.

Thus, in our example, we obtain from formula (28)

$$\underline{P}\{\tau > 1000\} = e^{-1} = 0.3679,$$

and from the precise formula (27) by taking logarithms, we obtain

$$\underline{P}\{\tau > 1000\} = 0.999^{1001} = 0.3673.$$

Let us calculate the mathematical expectation of the lifetime measured in terms of the number of landings. It is equal to

$$\underline{M}\{\tau\} = \sum_{r=1}^{\infty} (r - 1)\underline{P}\{\tau = r - 1\}. \tag{29}$$

By using (23) and making some simple calculations, we find

$$\underline{M}\{\tau\} = \frac{1-\gamma}{\gamma}. \tag{30}$$

If we remember that γ is usually small, we may write

$$\underline{M}\{\tau\} \approx \frac{1}{\gamma}.$$

In this case, the approximate formula (28) can be written in the form

$$\underline{P}\{\tau > K\} = \exp(-K/\underline{M}\{\tau\}) \ [1] \tag{31}$$

A geometric distribution of the lifetime τ corresponds to a situation in which the time is calculated in the form of discrete units (one landing, two landings, etc.). However, in the majority of problems, the time behaves like a continuous variable. The analogue of a geometric distribution for continuous life time τ is the exponential distribution. With an exponential distribution of the lifetime, the probability of the event that there will be no failures during the course of time T measured from the beginning of operation of the system or element, or, what amounts to the same thing, to the probability that the lifetime τ will exceed T if calculated in accordance with the formula

$$\underline{P}\{\tau > T\} = e^{-\lambda T}. \tag{32}$$

Remembering that

$$\underline{P}\{\tau \leq T\} + \underline{P}\{\tau > T\} = 1,$$

we have an expression for the distribution function

$$\underline{P}\{\tau \leq T\} = F(T) = 1 - e^{-\lambda T}. \tag{33}$$

In accordance with (5), the distribution density is, for $T \geq 0$,

$$f(T) = \frac{dF(T)}{dT} = \lambda e^{-\lambda T} \tag{34}$$

and $f(T) = 0$ for $T < 0$.

Fig. 3 shows graphs [2] of the change in the density $f(T)$ for differ-

---

[1] For simplification, fractional exponents e^x will be given as $\exp(x)$.

[2] In Fig. 3 and analogous drawings, the time T is calculated in arbitrary units (hours, hundreds of hours, etc.).

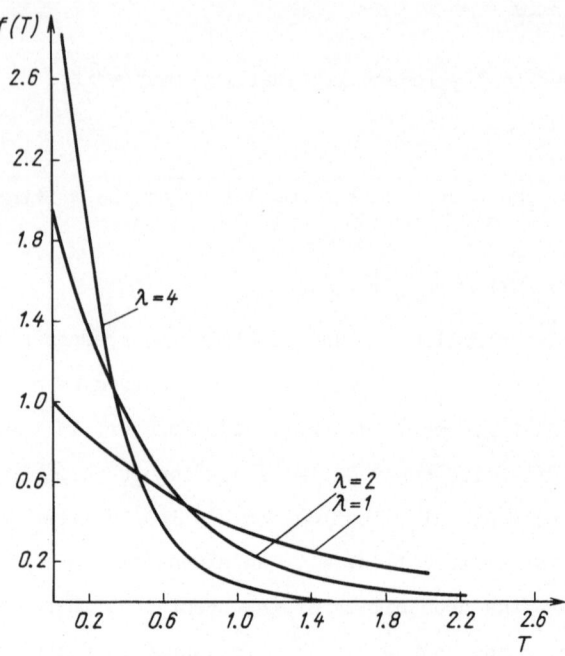

Fig. 3. The density of the exponential distribution for different values of the parameter λ.

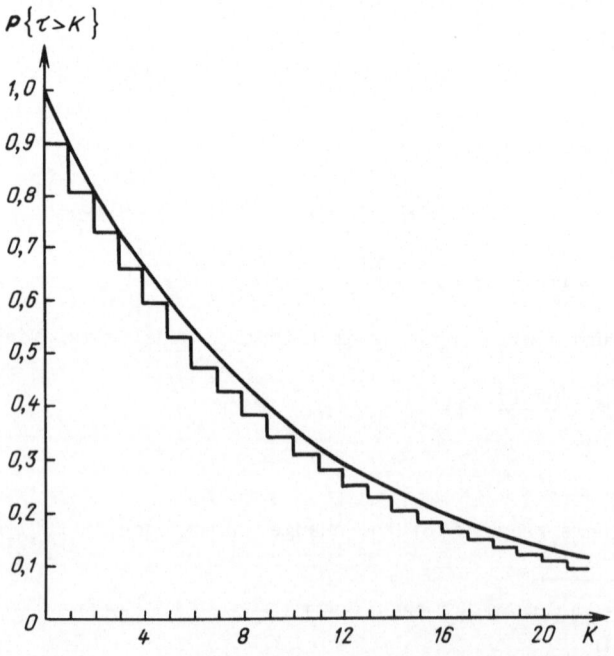

Fig. 4. The curves for the probability of failure-free operation for the geometric and exponential distributions: $\gamma = \lambda = 0.1$.

ent values of the parameter λ. Let us clarify the physical meaning of this parameter. The mean lifetime is

$$\underline{M}\{\tau\} = \int_0^\infty tf(t)dt = \lambda \int_0^\infty t\, e^{-\lambda t}\, dt = \frac{1}{\lambda}. \tag{35}$$

Thus, λ is the reciprocal of the mean lifetime. Now, we can write formula (32) as follows:

$$\underline{P}\{\tau > T\} = \exp(-T/M\{\tau\}) \tag{36}$$

This formula coincides with formula (31), which was obtained as an approximation to formula (27). This last formula is valid for a discrete scheme in which the lifetime was treated as the number of smooth landings. Figure 4 shows, for comparison, the graphs of the functions $\underline{P}\{\tau > T\} = R(T)$ calculated in accordance with (27) and (32) with the parameter λ taken equal to γ. The graph of the probability $P\{\tau > T\}$ as a function of T is usually called the reliability curve.

An estimate of the parameter λ on the basis of the data on the lifetime. Suppose that we have given the data on the lifetime of N units $\tau_1, \tau_2, \ldots, \tau_N$.
The first method of getting an estimate for λ consists in the following: We calculate the mean lifetime $\bar{\tau}$ from formula (1) and set the result equal to $\underline{M}\{\tau\} = 1/\lambda$ on the basis of (9). As a result, we obtain

$$\lambda = \frac{1}{\bar{\tau}} = \frac{N}{\sum\limits_{i=1}^{N} \tau_i}. \tag{37}$$

The second method consists in partitioning the set of initial data τ_i into two groups. One of these consists of the values τ_i that are less than some fixed number Θ. The other consists of all the remaining values τ_i. We denote the set of values of τ_i less than Θ by $m_1(\Theta)$. In accordance with formula (3), the ratio

$$\nu(\theta) = \frac{m_1(\theta)}{N}$$

is the accumulated frequency of failures. Therefore, beginning with (4), we may take

$$\underline{P}\{\tau \le \theta\} = F(\theta) = \nu(\theta).$$

Consequently, in accordance with (33),

$$1 - e^{-\lambda\theta} = \nu(\theta),$$

so that

$$\lambda = - \frac{\ln(1 - \nu(\Theta))}{\Theta} . \tag{38}$$

This formula enables us to estimate the parameter λ when we know the partition parameter Θ and the accumulated frequency corresponding to it.

The choice between the use of formula (37) or formula (38) to estimate λ depends on the specific conditions. If we are observing the values of τ_i for all N devices that are being tested (each device is tested up to the instant it fails), then formula (37) gives a more precise estimate of the parameter λ. However, a situation is also possible in which N devices that are involved in the operation are subjected to a test for some time Θ after the beginning of operation. Here, it is found that m_1 (Θ) of them are not functioning properly. Regarding a device that has failed, we can assert only that its lifetime τ_i is less than Θ; we can not say what the exact value of τ_i is. In this case, we can use formula (38) to obtain an estimate for the parameter λ.

Not infrequently, the time required to test all N units until they fail is too long and the tests must be terminated before the units fail. Such tests are called truncated tests. In particular, truncated tests are used in accordance with the total time of testing or the total number of failures recorded.

The book [11, Chapter III] expounds in detail the methods of estimating the parameter λ from the data of truncated tests.

Before turning to more complex schemes according to which failures can occur, let us stop again to look at the physical assumptions that lead to an exponential distribution of the time τ. As mentioned above, the lifetime is subordinate to an exponential distribution if the failures occur as a result of the action of "peak" loads. Let us look at this scheme of occurrence of a failure in greater detail. A load acting on the system as a whole and on its individual elements is always random. This is due to the inevitable random fluctuations in the external conditions and the inevitable random fluctuations in the interaction between the units.

Fig. 2 shows a random process of change of load acting on the wing of an airplane in horizontal flight. The load changes continuously and relatively smoothly. This means that, if we take two adjacent intervals of time (see Fig. 2, the intervals I and II), the values of the loads in these intervals will be connected with each other. Thus, if the load is small in the interval I, it is improbable that the load will be very high in the interval II adjacent to it. However, if we take the intervals of time I and III, which are separated by a fairly wide interval of time, the dependence of the value of the load in the interval III on the value in the interval I will be very slight whatever the value in the interval I was. Therefore, loads in intervals of time separated by a great distance from each other can be regarded as independent. We shall call this property of the load $S(t)$ the property of asymptotic independence (on the part of the value of the load $S(t_2)$ at the instant t_2) of the value of the load $S(t_1)$ at the instant t_1 when the difference $t_2 - t_1$ is great. The term "asymptotic" independence reflects the fact that the connection between the loads $S(t_1)$ and $S(t_2)$ decreases as the difference $t_2 - t_1$ increases.

Another peculiarity of the change in the load that is reflected in Fig. 2 is the fact that throughout the entire period of time in question the load does not have a directed change. This means that the "peak" values of the loads (indicated in Fig. 2 by the points 1, 2, ...) occur randomly and it is impossible to predict accurately the instant at which they will occur.

We shall refer to the absence of a directed change in the load $S(t)$ as the stationarity of the load. Thus, stationarity of the load acting on the wing of an airplane means that the airplane is flying over an area where random variations in rising and falling air currents constitute a random sequence that is homogeneous in time. Homogeneity will be ensured if the plane flies for a brief period of time when there is no change from night to day or day to night and the local relief over which it flies has no sharp changes (such as from dry land to ocean).

Every system is of finite "durability". Therefore, there is some limiting load S_1 that the system can take without failing. If the load $S(t)$ exceeds S_1, there will be an instantaneous failure. Fig. 2 shows conditionally the level S_1 which is the limit of admissible loads. The line $\bar{s}$ corresponds to the mean value of the load calculated as the arithmetic mean of the values of the loads that act during the course of an extended period of time. If the limiting load S_1 is much greater than the mean load $\bar{s}$, cases of the curve for the load $S(t)$ passing the level S_1 will be observed only very rarely. As a rule, the first crossing of the level S_1 will occur after a long interval of time τ. In view of the fact that the process of change in the load has the properties listed (asymptotic independence, stationarity), the time τ of first crossing of the level S_1 will be subordinate to an exponential distribution. Since the system fails immediately when the load exceeds S_1, the time τ is the lifetime of the system or unit.

In connection with this, we need to emphasize two points:

a) The level of maximum admissible load S_1 remains constant throughout the entire period of use of the system or unit.

b) A failure occurs not as a consequence of gradual change in the internal state of the units of the system but only as a consequence of an external random influence of a magnitude exceeding that which is admissible.

From this it is obvious that, in the case of an exponential distribution of lifetime, there is no point in resorting to preventive maintenance of the type of preliminary replacement of units or their periodic maintenance. Since failure occurs only as the result of an external influence, replacement of an old unit with a new one cannot affect the cause of failure. Still less can maintenance affect it. The natural way of increasing the reliability consists either in better construction of the system or unit or in decreasing the loads acting on it.

The threshold of sensitivity

Fig. 2 illustrates a situation in which the value S_1 of the maximum admissible load is independent of the time t of use of a device. Undoubtedly, this is an idealization of actual conditions. For example, the durability of any metallic construction, which is subject to aging and corrosion, decreases with the passage of time. In connection with our scheme, this means that the maximum admissible load does decrease gradually with the passing of time. The question is the degree to which the process of "weakening" of a system in time as a result of aging of its elements and whether it is sound policy to take this weakening into consideration when we are dealing with a restricted period of time of observation. Quite frequently, we encounter a situation in which the aging proceeds so slowly that it is virtually imperceptible during the course of a feasible interval of observation. This is the case with many constructions the proper functioning of which is extremely important, for example, bridges. In connection with this, cases are common when the weakening of the system takes places comparatively fast. More precisely, the observed weakening of the system proceeds up to an instant t* that is considerably less than the average length of service of the system itself: $t* < < \underline{M}\{\tau\}$. In this case, we must take into account this weakening process. In certain situations, this is done thus: We start with a conditional value of the time t_0 possessing the property that, for $t < t_0$, the system maintains its original properties and the maximum admissible load S_1 is so great that we can neglect the probability of failure in the interval $(0, t_0)$. After the instant t_0, a sharp decrease in the admissible load to a level S_1 ($< S_1$) occurs which, with detectable probability, can be exceeded by the operating load. Let us look at an example.

We know that the plastic and ceramic bases of electrical parts in radios are kept under special conditions in factories (vacuum, temperature, etc.) to eliminate moisture as much as possible and to maximize the electrostatic resistance. This ensures a high degree of resistance to "breakdown" at peak values of the voltage. However, in subsequent use, the material gradually becomes saturated with moisture. This saturation continues only to a certain point depending on the properties of the given object.

Increase in the moisture content leads to a considerable decrease in the resistance to "breakdown". Fig. 5 shows one possible course of change in the maximum admissible voltage S_1 depending on the time of use t, which corresponds to saturation of the base of the unit with moisture. In the interval I, when $S_1(t)$ is great, the probability that the operating voltage will cross the level $S_1(t)$ is extremely small and we can take it equal to zero. The interval II of sharp decrease in the value of $S_1(t)$ is relatively small in length and its influence on the general operating capacity of the unit is small. It is the interval III, when $S_1(t)$ remains approxi-

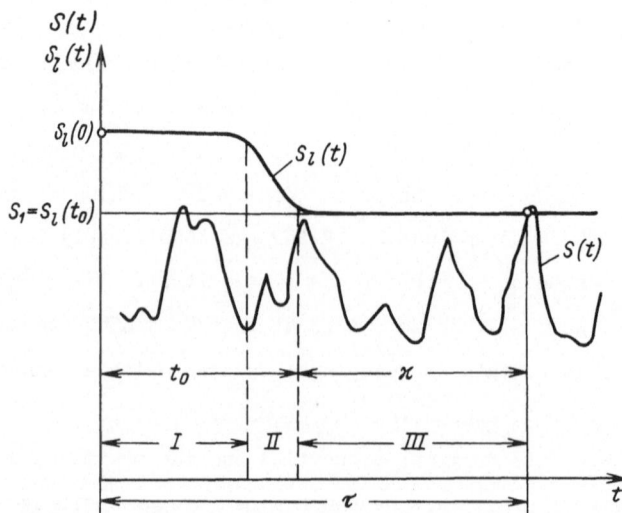

Fig. 5. The interaction of the operating stress S(t) and the breakdown stress $S_1(t)$.

mately constant and there is a real danger of breakdown, that is of greatest importance. The lifetime τ consists of two parts: the time t_0, which can be considered fixed and which corresponds to the guaranteed lifetime of the device, and the time $\varkappa$, which will have an exponential distribution.

We shall call the quantity t_0 the threshold of sensitivity. This name originated in metrology. Its significance, as far as the present problem is concerned, is that, during the time t_0, the device does not "feel" the effect of the load and it is only at the end of the time t_0 that this influence becomes perceptible and causes a danger of failure.

The corresponding distribution of lifetime is shown in Fig. 6. It differs from an ordinary exponential distribution by the fact that the entire curve is displaced to the right by an amount t_0. The density of the distribution is

$$f(T) = \begin{cases} \lambda e^{-\lambda(T-t_0)}, & T \geq t_0, \\ 0 & , T < t_0. \end{cases} \tag{39}$$

Here the quantities λ and T_0 are parameters of the distribution. The probability of failure-free operation is calculated in accordance with the usual rules except that we need to remember that the lifetime τ is always at least as great as t_0. Therefore,

$$\underline{P\{\tau > T\}} = \begin{cases} 1 & , \text{ if } T < t_0, \\ e^{-\lambda(T-t_0)}, & \text{ if } T \geq t_0. \end{cases} \tag{40}$$

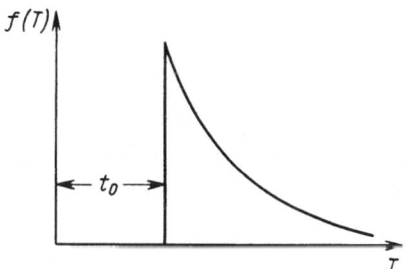

Fig. 6. The probability density of an exponential distribution with a threshold of sensitivity.

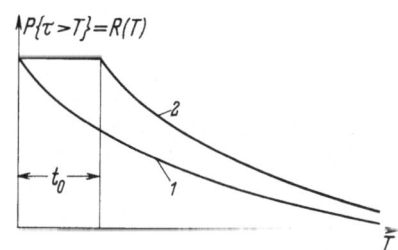

Fig. 7. The curves R(t) of one- and two-parameter exponential distributions.

21

Fig. 7 shows the curves R(t) corresponding to the distribution densities (34) [Curve 1] and (39) [Curve 2].

Let us refer to the distribution defined by the density (39) as a two-parameter exponential distribution.

We note that, on the basis of a two-parameter exponential distribution, we ignored the presence of the transitional interval II (see Fig. 5). The influence of this interval shows up in the fact that the form of the distribution curve (Fig. 8) differs somewhat from the form shown in Fig. 6. Specifically, there is a left branch of the distribution curve (see Fig. 8). This branch is hardly perceptible if the number of experimental data is small. Thus, a two-parameter exponential distribution should be regarded as a sort of substitute that gives a satisfactory description of the experimental data when we can neglect relatively small probabilities of failure as a result of the presence of the transitional interval of change in the maximum admissible load $S_1(t)$ (Fig. 5, the interval II).

The physical meaning of the threshold of sensitivity t_0 will be examined in section 3.

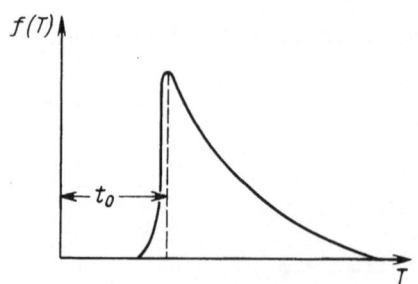

Fig. 8. A model of a two-parameter exponential distribution.

An estimate of the parameters λ and t_0 with the aid of exponential paper. Let us describe a simple practical device that enables us to get an approximation of the parameters of the distribution (39) without extensive calculations and that also enables us to judge how well the empirical distribution (that is, the cumulative frequencies) coincides with the distribution function. This device consists in using probabilistic paper of an exponential distribution (exponential paper[1]). Let us clarify with an example the rules for its use.

Example 2. The data on the lifetime of the divices and the accumulated frequencies are shown in Table 1. The lifetime has a two-parameter exponential distribution. Find an estimate for the parameters λ and t_0.
The value of $\nu(\tau_i)$ is taken equal to

$$\nu(\tau_i) = \frac{i - 1/2}{N}, \tag{41}$$

where N is the total number of data and i is an ordinal number. The quantities $\nu(\tau_i)$ are plotted onto the exponential paper described above (see Fig. 9). The quantity τ_i is the abscissa of the point and $\nu(\tau_i)$ is its ordinate. If we find $\tau_1, \tau_2, \ldots, \tau_N$ all equal, we need to lay off the arithmetic mean of the frequencies corresponding to the coincident values τ. Thus, in Table 1, we encounter $\tau_i = 44.0$ twice. Therefore, we plot $\nu(\tau_i) = (0.18 + 0.22)/2 = 0.20$. The scale for the ordinate is chosen in

Table 1. Data on the life time of devices.

Lifetime τ_i (hours)	Accumulated frequency $\nu(\tau_i) = \left(i - \frac{1}{2}\right)\frac{1}{N}$	Lifetime τ_i (hours)	Accumulated frequency $\nu(\tau_i) = \left(i - \frac{1}{2}\right)\frac{1}{N}$
36.0	0.02	63.5	0.54
38.5	0.06	69.0	0.58
41.0	0.10	72.0	0.62
42.5	0.14	75.0	0.66
44.0	0.18	81.0	0.70
44.0	0.22	86.5	0.74
48.0	0.26	92.0	0.78
49.0	0.30	100.5	0.82
51.0	0.34	107.5	0.86
53.0	0.38	120.0	0.90
56.0	0.42	156.0	0.94
60.0	0.46	156.0	0.98
61.0	0.50		

---

[1] Exponential paper is one form of probability paper. One can read further about it in the book by A. H a l d [31, Chapter VI]. There are various types of probability paper for different frequently encountered distributions.

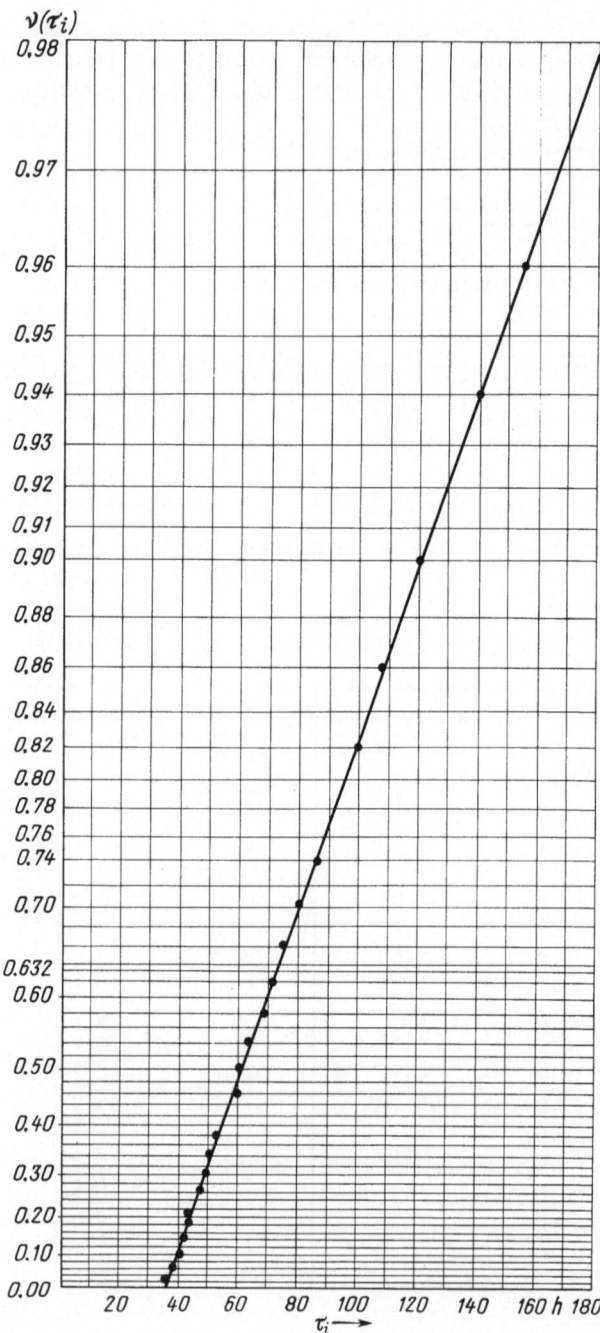

Fig. 9. Plotting of the data of Table 1 onto probability paper for an exponential distribution.

such a way that the accumulated frequencies $\nu(\tau_i)$ will lie on a straight line when the τ_i have an exact exponential distribution. We can draw a straight line approximately through the points obtained. In doing this, we should try to have about as many points above as below the line. The degree of closeness of all the points to a straight line testifies to the degree of compatibility of the initial data with the hypothesis of an exponential distribution. The agreement of the data of Table 1 is extremely good. The parameters t_0 and λ are approximated from the graph of Fig. 9 in the following way. For t_0 we take the abscissa of the point of intersection of the straight line with the horizontal axis. In the present case, $t_0 = 36$. We then find the value τ corresponding to the ordinate 0.632. Let us denote it by τ^*. In the present case $\tau^* = 73$. An estimate of the parameter λ is made in accordance with the formula

$$\lambda = \frac{1}{\tau^* - t_0} = 0.027. \tag{42}$$

Superposition of exponential distributions

In section 1, we showed that failures resulting from peak loads are typical when there are construction defects. Here, we assume that all exemplaires are homogeneous in quality.

On the other hand, if there are technological defects in the units, this portion of the units will have lowered reliability. Fairly typical is the case when the main portion of units is of high quality and correspondingly high reliability but a certain small portion of them have technological defects. The influence of peak loads on the first portion is weak, whereas the influence of the same loads on the remaining devices will be noticeable. Let $(1 - \varepsilon)$ denote the portion of high-quality units and let ε denote the portion of defective ones.

If from the total set of units we choose a unit at random, it will belong to the first portion with probability $(1 - \varepsilon)$ and it will belong to the second portion with probability ε. One can easily see that the probability of failure-free operation for a period of time T is, for a randomly chosen unit,

$$R(T) = (1 - \varepsilon)R_1(T) + \varepsilon R_2(T), \tag{43}$$

where $R_1(T)$ is the probability of failure-free operation of a "good" unit and $R_2(T)$ is the probability of failure-free operation of a "bad" one.

In accordance with (43), the density of the distribution of the lifetime is represented in the form

$$f(T) = \begin{cases} (1-\varepsilon)\lambda_1 e^{-\lambda_1 T} + \varepsilon\lambda_2 e^{-\lambda_2 T}, & T \geq 0, \\ 0 & , T < 0. \end{cases} \qquad (44)$$

where λ_2 is the parameter of the units with defects and λ_1 is the parameter of units without defects.

The density (44) differs from the usual curve for the density of an exponential distribution with the same value of the mathematical expectation in that it has a steeper descent and the "tail" of the curve (see Fig. 10) extends somewhat further on the side of large values.

When there is a properly organized system for assembling the information regarding failures on the basis of experimental data, we can derive estimates for the parameters ε, λ_1, and λ_2. Thus, suppose that we know that, out of N units that have failed, N_1 failed because of a technological defect. From this, we see that $\varepsilon = N_1/N$.

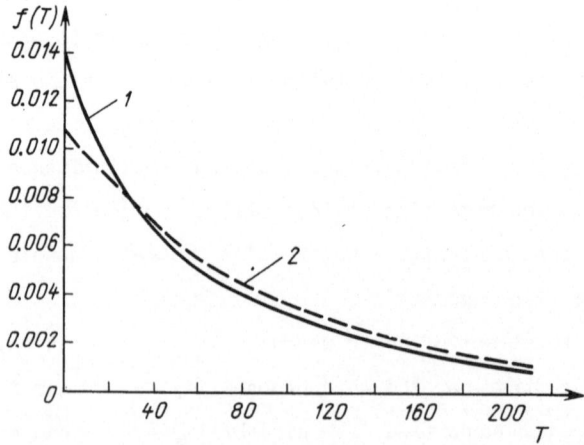

Fig. 10. Density of distribution of the lifetime in accordance with (44) [Curve 1] and in accordance with (34) [Curve 2] with $\varepsilon = 0.1, \lambda_1 = 0.01, \lambda_2 = 0.05$, and $\lambda = 0.0109$. The mathematical expectations of the two distributions coincide.

If we break the data regarding the lifetime into two subgroups $(\tau_1, \tau_2, \ldots \tau_{N_1})$ and $(\tau'_1, \tau'_2, \ldots, \tau'_{N-N_1})$ for both groups of units, we find the parameters λ_1 and λ_2 from the formulas

$$\lambda_1 = \frac{N - N_1}{\sum\limits_{i=1}^{N-N_1} \tau'_1} \,, \tag{45}$$

$$\lambda_2 = \frac{N_1}{\sum\limits_{j=1}^{N_1} \tau_j} \,. \tag{46}$$

We assume that we do not find defects of both types, technological and constructional, in a single unit. Although, in actuality, this is not the case, use of this assumption and of the proposed method of estimating the parameters does not lead to significant errors if the technological defect causes a considerable decrease in the reliability of the unit. The case of simultaneous action of two causes of failures will be considered in section 5.

Chapter III

Cumulative Damages

Models of wear

Models of wear. In Chapter I, we noted that, among the reasons for
failures, an important place is occupied by the aging of a system.
No matter how perfect the construction of a system and its elements or the technology of their production, the material of which
the units are made and combinations of them will, with the passage
of time, undergo irreversible changes. These changes are caused by
corrosion, abrasion, accumulation of deformations, fatigue, diffusion of one material in another, etc. In a single system or even
in a single unit, these processes are superimposed on each other,
they interact with each other and, in the final analysis, they
cause a change in the operating characteristics.

A simple example is the change in the space between a shaft
and bearings that rub against each other. As the bearings rotate
in conjunction with each other, various complicated processes
take place: oxidation of the surface layers, hardening and accumulation of fatigue in them, abrasive cutting, and sticking as a
result of adhesion. These processes together lead to a gradual increase in the gap and loss of functioning capacity of the bearing.
In the use of receiving-amplifier tubes, one observes a gradual
liberation of gases (oxygen, aqueous vapors) and their accumulation in the bulb of the tube, penetration of the oxide layer of
the cathode as the result of diffusion processes, evaporation of
barium, etc. All this is reflected in the parameters of the lamp,
for example, in the steepness of the characteristic and the impulse current of the cathode, which gradually decrease as a result
of the processes mentioned (see Fig. 11) [24].

For a system as a whole and its elements to be capable of function-

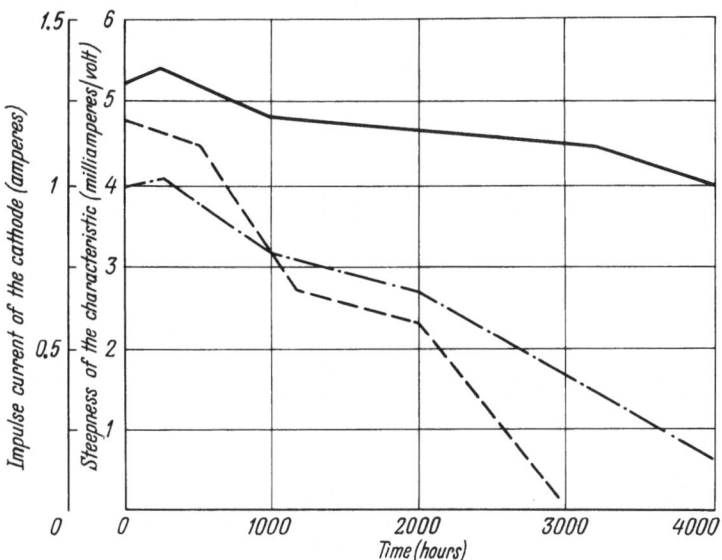

Fig. 11. Change in the parameters of the 6Zh1P lamps in the course of protracted use. The dashed line denotes the impulse current of the cathode. The dot-and-dash line denotes the steepness of the characteristic when the voltage across the filter is 5.7 V. The solid line at the stop denotes the steepness of the characteristic when this voltage is 6.3 V.

ing, their operating characteristics must lie between certain limits defined by the form and purpose of the system. Thus, the gap in a coupling must lie between limits ensuring smooth motion and the absence of dry friction. The steepness of the characteristic of the lamp must not be less than some particular quantity and so forth. When the operating characteristic gets beyond the prescribed limits, the unit or system begins to operate unsatisfactorily, and this qualifies as a failure.

The operating characteristic does not need to change monotonically under the influece of aging although in the majority of cases it does so (see Fig. 12). The gap in a coupling, the steepness of the characteristic, etc., change monotonically. Our further reasoning will be based on the assumption that the change in the operating characteristic is monotonic. We also assume that a failure occurs as soon as the operating characteristic gets beyond the limit given by the technological conditions. The lifetime τ is determined by the instant the operating characteristic gets beyond the limiting level.

We denote by $\eta(t)$ the value of the operating characteristic at the instant t. Changes in $\eta(t)$ with the passage of time are conditioned both by external factors and by the course of physical processes that take place inside the object. Furthermore, the form of the sample function of $\eta(t)$ depends on the initial state of the object, for example, the quality of the manufacturing.

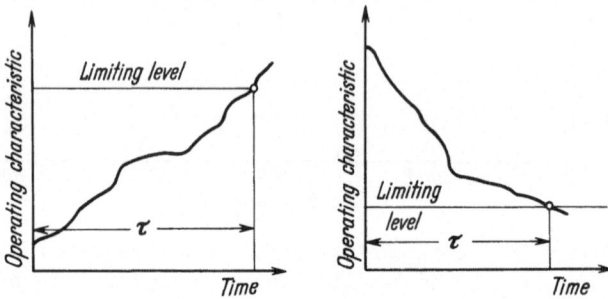

Fig. 12. Determination of the lifetime in the case of monotonic change in the operating characteristic.

The simplest assumption regarding the change in $\eta(t)$ is that it changes linearly (see Fig. 13). The slope of the line depends on the initial state of the object. All "weak" copies have a larger angle of inclination of the operating characteristic.

For each unit, the change in $\eta(t)$ has a completely determined, non-

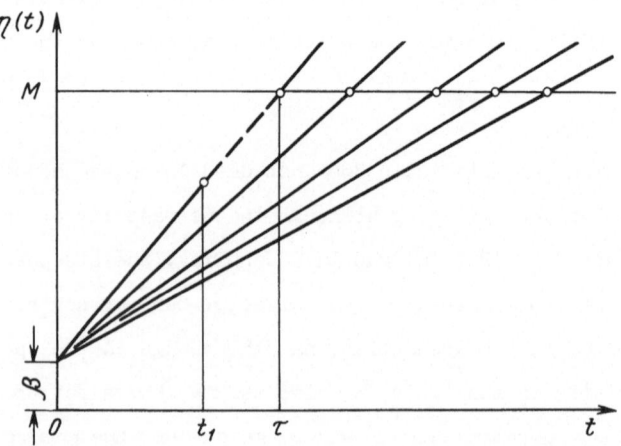

Fig. 13. Linear sample functions of the wear $\eta(t)$; M denotes the maximum admissible level of wear.

random nature. To determine the behavior of $\eta(t)$ for a given unit during the interval (t_1,τ), we need only draw a straight line through the points $\eta(0)$ and $\eta(t_1)$, as shown in Fig. 13. The randomness in the change $\eta(t)$ consists in the fact that the coefficient α in the equation describing $\eta(t)$

$$\eta(t) = \alpha t + \beta, \tag{47}$$

is a random variable determined by the initial state of the object. Obviously, α is the rate of change of $\eta(t)$:

$$\alpha = \frac{d\eta(t)}{dt} = \xi(t). \tag{48}$$

Suppose that a failure occurs when

$$\eta(t) \geq M.$$

In this case, the lifetime of τ is determined from the formula

$$\tau = \frac{M - \beta}{\alpha} . \tag{49}$$

Fig. 14 shows the dependence of the wearing of automobile tires on

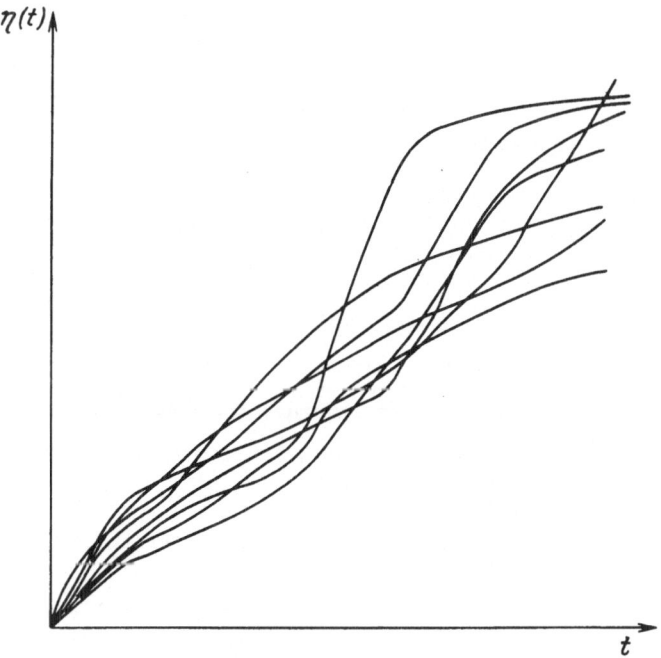

Fig. 14. Sample functions of the wear of tires as a function of the time of use (length of flight).

their length of use. [1] These curves cannot be represented in the
form of the straight lines shown in Fig. 13. It is typical of these
curves for the wear on tires that they are closely interwoven and
that the path of an individual curve for this wearing is a sinuous
curve. This is caused by the conditions under which the tire is
used. The wear on the tire is greatest on rough roads. The use of
brakes has an effect on the wear, and this use is caused by random
situations on the road. Therefore, the rate of wear of an individual tire changes continually and these changes are of a random
nature. Consequently, we cannot extrapolate definitely the curve
representing the wear on an individual tire in the future from the
results of observation during the interval $(0, t_1)$, as was done
above (see Fig. 13). We can only make a probabilistic judgment as
to its future behavior.

When the rate of wear is subject to random changes, we cannot write
a formula for $\eta(t)$ similar to formula (47) [2] or derive a simple
relation for the determination of the lifetime of the type (49).
These situations, illustrated in Figures 13 and 14, seem contradictory. Whereas the future behavior of the sample function of
$\eta(t)$ in Fig. 13 was completely determined by its past behavior,
then the future behavior of the sample function of $\eta(t)$ in Fig. 14
is almost independent of its past behavior. The situations described are not encountered very often in their pure form. In
practice, elements of wear of both types always exist in practice.
For example, suppose that we have two tires, the quality of one of
which is good and the other poor. The sample functions of the wear
of the two tires will both be sinuous but, nonetheless, the poor
tire will wear out faster on the average than the good one. This
is shown in Fig. 15, which reflects a situation midway between
those of Figures 13 and 14.

Let us consider a few cases encountered in practice.

---

[1] From the data of Y e. F. N e p o m n y a s h c h i y .

[2] $\eta(t)$ can be represented in the form of an infinite series with
random coefficients. The formal writing of such a series is not
of great value in practical computations of the reliability.

Sometimes, most frequently in the course of laboratory (bench) tests of units, the properties (moisture, temperature, rules of switching, etc.) of the external medium are maintained at a constant level. This leads to the fact that the influence of the external factors on all the units tested is the same and all exemplaires display a constant rate of wear (aging). The only difference between the units is that the initial value of the operating parameter $\eta(0)$ varies from one to the other and is a random variable.

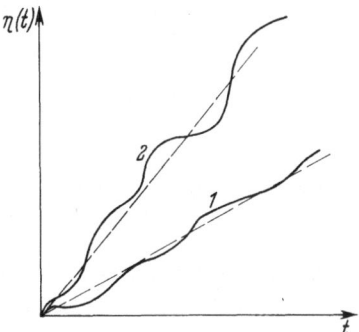

Fig. 15. Sample functions of the wear on tires of good (1) and poor (2) quality.

In this case, the sample functions of the process of change in the operating characteristic take the form shown in Fig. 16. Such a picture, which can almost be called "ruled", is observed especially often on bench tests of vacuum tubes and transistors. We may expect that these same units, tested under different conditions of use (for example, each unit may have its own temperature regime and its own frequency of being turned on), would behave in different ways and that it would disclose differing rates of change in the operating parameter (see Fig. 17).

In practice, we also encounter situations in which the behavior of the sample function of $\eta(t)$ is subject to strong variations although the external load is quite stable. This is the case, for example, with the accumulation of corrosion, aging, and creep of metals. A common feature of these processes is a high complexity of the medium in which these processes take place. A metal is a polycrystal and consists of grains of varying forms and orientations

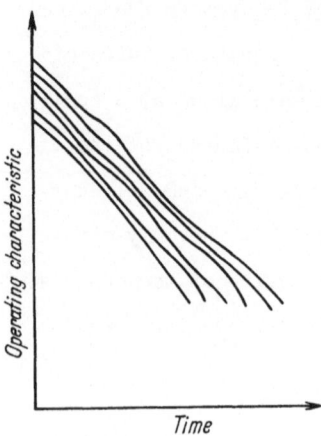

Fig. 16. The case in which the rate of change in the operating characteristic is approximately constant for all units.

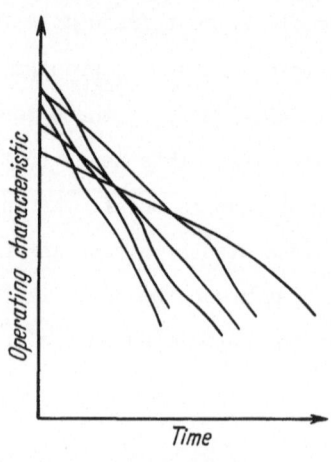

Fig. 17. The case in which the rate of change of the operating characteristic differs from one unit to the next.

on the boundaries of which various chemical elements are concentrated. Cavities, cracks, etc., are formed inside the metal. Therefore, the properties of the metal are subject to random variations from one portion of the metal to another. [1]

Let us look at the process of propagation of corrosion inside a metal. The metal consists of elementary volumes the properties of which have random variations. In particular, their resistivity to penetration of corrosion varies randomly. Corrosion begins at the surface (point corrosion) and proceeds from volume to volume. When corrosion reaches a given region in the metal, it acquires a certain rate of propagation depending on the properties of that region. Since these properties are random, the rate of corrosion is also random. This rate of corrosion will undergo random changes as it proceeds from one region to the next. Thus, although the external conditions (the corroding medium) remain unchanged, the rate of wear (rate of corrosion) is subject to random variations. If

---

[1] Systems in which the elements or elementary volumes forming them possess random properties are studied in statistical mechanics. Similar approaches are used in the study of metals.

the initial properties of objects made of metal do not differ too strongly, then the curves for the accumulation of corrosion (curves for the corrosional wear) will resemble the sample functions shown in Fig. 14.

With regard to the process described by equation (47), the rate of wear remains constant for each individual object throughout the period of its use. Changes in the rate of wear $\xi(t)$ are observed only when we turn from one object to another. Fig. 18 a shows the sample functions of the rate of wear $\xi(t)$ for several objects L_1-L_4.

The sample functions of the rate of wear of tires appear quite differently (see Fig. 18 b). Here, we have random variations in the rate about some constant level characterizing the mean rate of wear. The absence of a single sample function (change in the rate from one tire to another) is not of a systematic nature. This means that, no matter what sample function we use to estimate the mean rate of wear or any other characteristic, the data obtained in this way will not have significant differences. The sample functions of the rate that are shown in Fig. 18 b have yet another important peculiarity. If we look at the two intervals of time (T_1, T_2) and (T_3, T_4) and estimate the characteristics of the wear in each of them, the values obtained will differ only in the random sense. Thus, the estimate of the average velocity of wear on each of these two intervals yields approximately the same values.

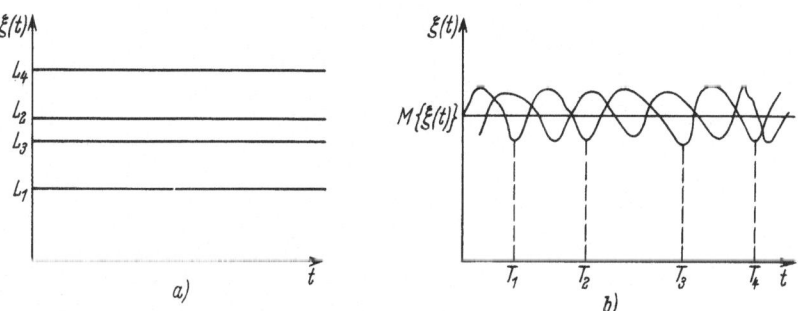

Fig. 18. A case in which the rate of wear is constant for each unit (a) and variable (b) (on the average, the rate of wear is the same for all units).

From a formal standpoint, the process of change in the rate $\xi(t)$ in the course of time can be regarded as a random process. The properties described above mean that this process is stationary and the connection between the quantities $\xi(t_1)$ and $\xi(t_2)$ decreases with increase in the difference $t_2 - t_1$.

Physically, these processes bear witness to the fact that all the objects are of the same quality and the properties of the objects do not change appreciably until the wear reaches its limiting value.

If the initial quality of the objects changes appreciably, then each of them will have its own quality of average rate of wear. The sample functions of the velocity $\xi(t)$ will differ systematically from each other according to their mean levels.

For by no means all objects does the mean rate of wear remain constant in time. For example, the mean rate of accumulation of aging in metals [21] decreases with the passage of time. Another example is the gradual decrease in the velocity of diffusion of one metal in another as the surface layers become saturated. The increase in the velocity of wear can take place under the influence of a gradual change in the conditions of operation of the given unit. Similarly, with the accumulation of aqueous vapors in the bulb of a vacuum tube, the rate of oxidation of the anode and cathode increases.

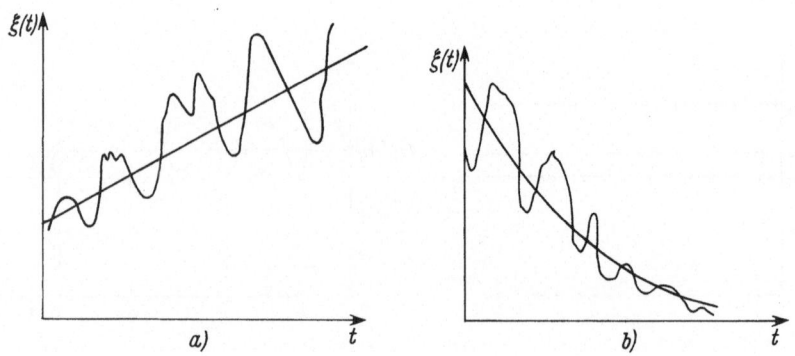

Fig. 19. The case in which the rate of wear increases linearly on the average (a) and decreases in accordance with formula (52) (b).

A rather general description of the behavior of the rate of wear $\xi(t)$ can be given in the form

$$\xi(t) = V(t) + v(t)\rho(t), \tag{50}$$

where $V(t)$ and $v(t)$ are certain determined functions and $\rho(t)$ is a stationary random process.

In a large number of cases, it will be sufficient to represent the rate in the form

$$\xi(t) = v(t)\rho(t). \tag{51}$$

Thus, if $v(t) = at + b$, then the rate will increase linearly on the average as is shown in Fig. 19 a; if

$$v(t) = \frac{a}{b+t}, \tag{52}$$

then the rate will decrease on the average in a manner similar to that represented in Fig. 19 b.

Without loss of generality, we may assume that the mathematical expectation (that is, the mean value) of the process $\xi(t)$ is a constant quantity and is equal to unity; that is,

$$\underline{M}\{\rho(t)\} = 1. \tag{53}$$

It follows from formula (51) that the mathematical expectation of the velocity $\xi(t)$ is then given by the equation

$$\underline{M}\{\xi(t)\} = v(t). \tag{54}$$

We may assume that

$$\underline{D}\{\rho(t)\} = 1. \tag{55}$$

Then, the variance

$$\underline{D}\{\xi(t)\} = v^2(t). \tag{56}$$

It follows from the last equation that, if the mean rate $v(t)$ increases, then the variance in the rate also increases. In Fig.19a, this is reflected in the ever-increasing spread of the random oscillations in the rate $\xi(t)$. If $v(t)$ decreases, then the variance $\underline{D}\{\xi(t)\}$ also decreases. Fig. 19b shows the damping of the random variations in the rate $\xi(t)$.

Since the process of wear is irreversible, we always have

$$\xi(t) \geq 0, \tag{57}$$

with equality holding, if at all, only at individual points.

The amount of wear that has occurred up to the instant t is the integral of the rate of wear and hence it can be written in the form

$$\eta(t) = \int_0^t \xi(x)dx. \tag{58}$$

The mathematical expectation $\underline{M}\{\eta(t)\}$ of the wear is represented as follows:

$$\underline{M}\{\eta(t) = \int_0^t \underline{M}\{\xi(x)\}dx = \int_0^t v(x)dx. \tag{59}$$

In particular, if the rate of wear is given by equation (52), then the mathematical expectation of the wear will be given by the equation

$$M\{\eta(t)\} = a[\ln(b+t) - \ln b], \tag{60}$$

that is, the wear increases on the average as the logarithm of the time.

In what follows, we shall consider successively idealized models of processes of change in the values of the operating parameters, taking various situations into account that arise with real objects. Here, we shall pay especial attention to the physical treatment of the schemes in question. We shall call all these models models of wear independently of their physical nature. In so doing, we shall use the word wear in the broad sense as an irreversible process of change in an operating parameter.

The gamma distribution of the lifetime

Let us consider an idealized scheme of wear with the following properties: the mean rate wear is constant; the initial quality of the objects is the same for all of them; the rate of wear is subject to random variations.

The situation when the first damage leads to failure of the object was considered in Chapter II. A natural generalization of this situation is one in which an accumulation of several injuries is necessary for failure of the object. For example, if the wear on a tire occurs only as the result of braking, then several instances of braking must occur for the maximum admissible value of the wear to be reached; that is, several injuries must occur.

Suppose that isolated injuries of equal magnitude occur at random instants of time and that, after a total of r injuries have occurred, the object gets out of operation (a failure occurs). The injury consists in the fact that the wear increases stepwise by an amount y, which is constant. Fig. 20 shows the scheme for the accumulation of injuries. The dashed line corresponds to the mean value of wear that has occurred up to the instant t. The random variations in the value $\eta(t)$ of the wear about this straight line are conditioned by the randomness of the instant of occurrence of a step increase in the wear.

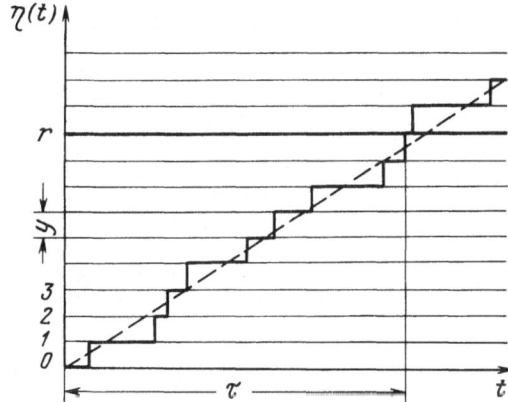

Fig. 20. Sample function of a process of accumulation of injuries.

It is assumed that the probability of occurrence of a jump in the wear during the time from T to $T + \Delta_T$ is equal to

$$\gamma(T) = \gamma = \lambda \Delta_T + o(\Delta_T) \tag{61}$$

and is independent of the number of jumps (the number of injuries) during the time from 0 to T. We thus assume that the probability of each subsequent injury is independent of the number of injuries already sustained. Let us pause to look in greater detail at the significance of this assumption.

A wear process always has at least three zones (see Fig. 21). The zone I is called the running period. In the process of accommodation, directed changes take place in the object; the object accommodates itself, as it were, to the loading conditions. For example,

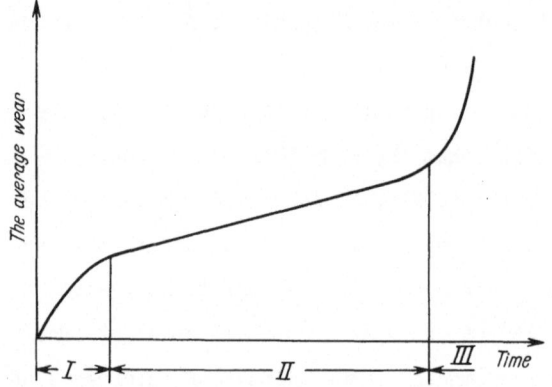

Fig. 21. A typical wear curve.

in the process of accommodation of rubbing a pair of roller bearings, a directed change in the roughness of the surface takes place. The rate of wear decreases constantly during this time. A characteristic feature of this accommodation is the mutual dependence of the increments in the wear. Fig. 22, shows two increments in the wear: the increment $\Delta\eta_1$ during the period from T_1 to $T_1 + \Delta_T$ and the increment $\Delta\eta_2$ during the period from T_2 to $T_2 + \Delta_T$. If the increment $\Delta\eta_1$ is great, this means that during the interval

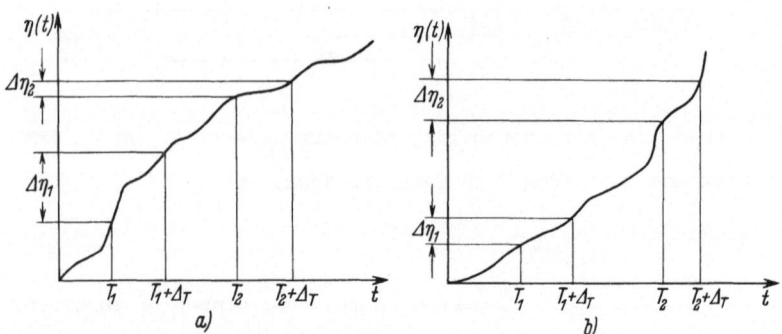

Fig. 22. Sample function of a process of wear in the zone of accommodation (a) and in the zone of catastrophic wear (b).

40

$(T_1, T_1 + \Delta_T)$ there was an intensive accommodation and then it is probable that during a subsequent interval $(T_2, T_2 + \Delta_T)$ the intensity of accommodation will decrease: that is, it is probable that the value of $\Delta\eta_2$ will prove comparatively small. Therefore, in the first zone, the increment in the wear during the interval from T to $T + \Delta_T$ depends on what is value was in the preceding instant of time.

Zone II (see Fig. 21) is called the zone of steady-state or normal wear. In this zone, the object acquires certain relatively stable properties corresponding to the loading conditions. The zone of normal wear occupies the greatest portion of the time of functioning of the object. With regard to a friction pair consisting of a shaft and a set of bearings that rub against it, the zone of normal wear is characterized by constancy of relief in the roughness and by the constant increase in the gap without change in the physical picture of the interaction of the roller with the bearing. In this zone, the value of the increment $\Delta\eta_1$ does not affect appreciably the value of the increment $\Delta\eta_2$.

Zone III (see Fig. 21) is called the zone of catastrophic wear. In the zone of normal wear, the changes that take place in the object are primarily of a quantitative nature. But when the wear reaches a certain value, there occurs a qualitative jump in the state of the object consisting in a significant change in the physical picture of the phenomena that are occurring. New factors, not previously discernible begin to act on the rate of wear, and this leads to a worsening of the state of the object and its destruction. The zone in which a sharp increase in the wear as the result of a sharp change in the physical picture of this wear is called the zone of catastrophic wear. When this occurs, the changes in the state of the objects, just as in the accommodating zone, are of a directed nature and the increments in the wear in this zone are mutually dependent. Here, large increments in the interval of time from T_1 to $T_1 + \Delta_T$ (see Fig. 22b) cause even larger increments in the interval from T_2 to $T_2 + \Delta_T$. For a rubbing shaft-and-bearings combination, a catastrophic wear occurs when there is a wide gap due to loads in the form of blows and a violation of the lubricating condition. From the instant catastrophic wear begins, there is a directed decrease in the quality of the surface and an increase in the roughness. There ensues sticking with subsequent tearing off of large pieces and finally a destruction of the bearing.

It follows form the description of the zones of wear that the assumption that the probability of damage is independent of the number of injuries already sustained corresponds to the zone of normal wear.

Let us show now that equation (61) ensures constancy of the mean rate of wear. Let us denote by $X(T)$ the number of injuries sustained up to the instant T. Then, the amount of wear at the instant T is

$$\eta(T) = yX(T). \tag{62}$$

The difference

$$\eta(T + \Delta_T) - \eta(T) = y[X(T + \Delta_T) - X(T)] \tag{63}$$

is the increment in the value of the wear during the time Δ_T. Let us calculate the mathematical expectation of this difference. We note that, during the time Δ_T, the wear can, in accordance with equation (61), either undergo an increase equal to y (this with probability γ) or undergo an increase equal to zero (this with probability $1 - \gamma$). Therefore, the mathematical expectation of the increase in wear is equal to

$$\underline{M}\{\eta(T + \Delta_T) - \eta(T)\} = \gamma y + (1 - \gamma)0 = [\lambda\Delta_T + o(\Delta_T)]y. \tag{64}$$

Remembering that the mathematical expectation of the difference between random variables is equal to the difference between their mathematical expactations, we obtain

$$\underline{M}\{\eta(T + \Delta_T)\} - \underline{M}\{\eta(T)\} = [\lambda\Delta_T + o(\Delta_T)]y. \tag{65}$$

If we divide both sides of this equation by Δ_T and take the limit as $\Delta_T \to 0$, we obtain

$$\frac{d\underline{M}\{\eta(T)\}}{dT} = \lambda y. \tag{66}$$

42

From this it follows that λy is the mean rate of wear. Since λy
is independent of the time, the mean rate is constant.

Now, we are in a position to assert that the model that we have
been describing of a process of wear corresponds to a situation
in which the accommodation is already completed and a catastrophic
wear has not yet occurred. The rate of wear is constant on the
average and the initial quality of all objects is the same.

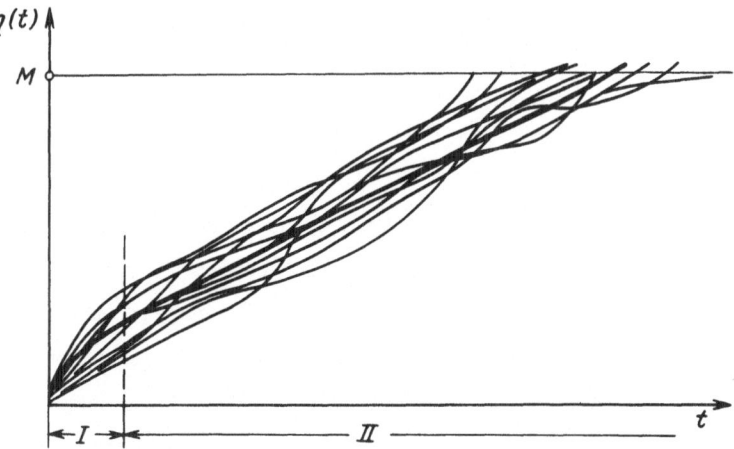

Fig. 23. Sample functions of a wear process. Here, M denotes the
maximum admissible level of wear, I denotes the zone of accommodation, and II denotes of normal wear.

Certain judgements can be made regarding the acceptability of the
description proposed for a specific object on the basis of the external form of the sample functions. Fig. 23 shows functions of a
wear process with the following important features: the interval
of accommodation is small in comparison with the interval of normal wear; the sample functions extend out to the maximum admissible level M of wear and tear before catastrophic wear occurs;
once the interval of accommodation is passed, the sample functions
become interwoven. After the accommodation, the rate of wear is
on the average constant. As a rule, sample functions of the wear
of objects of high quality possess these features. When sample
functions of a wear process are of the type shown in Fig. 23, we
can safely assume that the wear process corresponds to the scheme
of accumulated injuries given by equation (61).

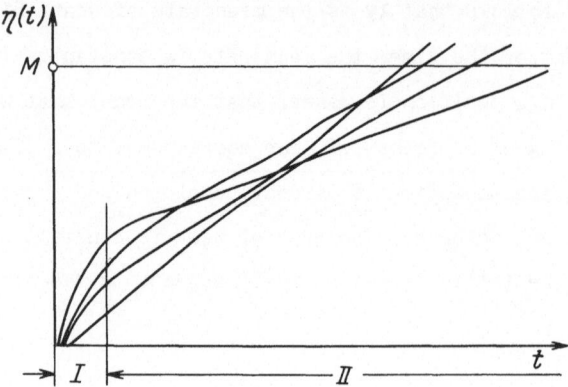

Fig. 24. The case in which there is only a slight amount of interweaving of the sample functions in the zone of normal wear.

The sample functions shown in Fig. 24 appear quite different. Although the zone of accommodation still occupies only an insignificant portion of the functioning time of the object, the rate of
wear is constant on the average, and there is no catastrophic wear,
nonetheless the wear scheme given by equation (61) is unsuitable.
The point is that after the accommodation the sample functions of
the wear become separated and their subsequent behavior depends
exclusively on the initial quality of the objects, and the rate of
wear of an individual unit (compare with Fig. 23) is not subject
to random variations.

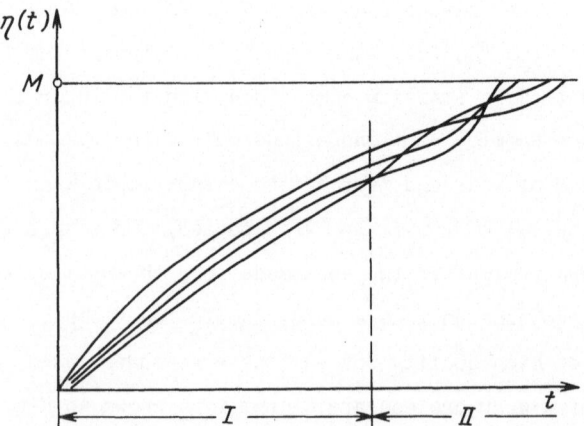

Fig. 25. The case in which the interval of accommodation occupies
the main portion of the operating time.

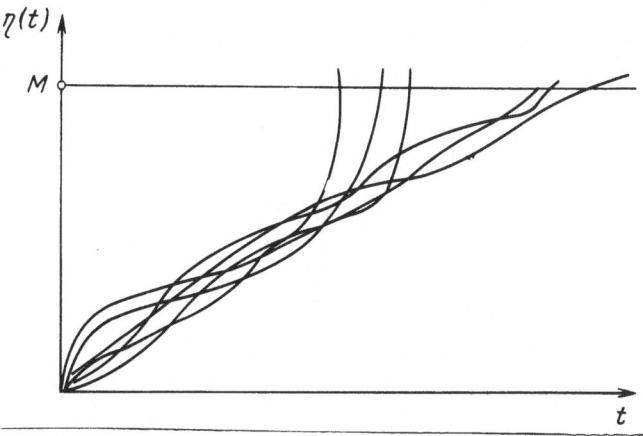

Fig. 26. Sample functions reflecting the case in which catastrophic wear occurs before the maximum admissible level M is reached.

The description given of a process of wear is also inappropriate for the situations shown in Fig. 25 and 26. Fig. 25 shows the case in which the zone of accommmodation occupies almost the entire time of functioning of the object. Fig. 26 shows a case in which catastrophic wear occurs in the case of individual objects before the maximum admissible level of wear is reached.

At the end of this chapter, in the section entitled "Analysis of sample functions of wear", we shall present methods of analysis that are based on objective estimates and not on visual analysis. With the aid of these methods, we can obtain quantitative estimates as to how close the actual picture of the wear is to the idealized scheme considered in the present section.

In those cases when it is impossible to observe the sample functions of the wear, a judgement as to the acceptability of the scheme of accumulated injuries must be based on an analysis of the physical picture of the process of wear. The chief attention should be concentrated on the following three points: homogeneity of the objects, variation in the loading conditions, protractedness of the accommodation.

Obviously, the scheme of accumulated injuries is suitable when the mass production ensures a high degree of homogeneity in the initial quality of the objects (a high degree of homogeneity in the original materials, a stable technological process, strict quality control); weights acting on objects in the course of their use vary between rather wide limits; the accommodation is to some extent ensured under factory conditions and, in the course of use, occupies an insignificant amount of time.

A scheme of accumulated injuries corresponds to a gamma distribution of the lifetime τ. The density of this distribution has the form

$$f(T) = \begin{cases} \dfrac{1}{\Gamma(r)} \lambda^r T^{r-1} e^{-\lambda T}, & T \geq 0, \\ \qquad\quad 0 & , \ T < 0. \end{cases} \qquad (67)$$

Here, r is the number of injuries necessary for a failure to occur and $\Gamma(r)$ is the gamma function defined by

$$\Gamma(r) = \int_0^\infty x^{r-1} e^{-x} dx. \qquad (68)$$

For integral values of r, we have the relation

$$\Gamma(r) = (r-1)! \qquad (69)$$

Let M denote the maximum admissible level of wear; that is, suppose that a failure occurs whenever $\eta(t) \geq M$. It then follows that the number of injuries up to a failure is given by the ratio

$$r = \frac{M}{y}. \qquad (70)$$

The quantity λ gives the average rate of wear and, in accordance with (66), we have

$$\lambda = \frac{1}{y} \frac{dM\{\eta(T)\}}{dT}. \qquad (71)$$

The distribution function F(T) of the gamma distribution is given in accordance with (67) by

$$F(T) = \int_0^T f(t)dt = \lambda^r / \Gamma(r) \int_0^T t^{r-1} e^{-\lambda t} dt. \qquad (72)$$

For integral values of r, we integrate (72) by parts and obtain

$$F(T) = 1 - \sum_{k=0}^{r-1} (\lambda T)^k / k! \ e^{-\lambda T}. \qquad (73)$$

For r = 1, the density (67) coincides with the density of the exponential distribution, which agrees with the model of the occurring of an exponential distribution when there is a single injury. The curves of the density (67) are shown in Fig. 27. We note that, for small values of r, these curves are asymmetric and that, with increasing r, they become more and more symmetric.

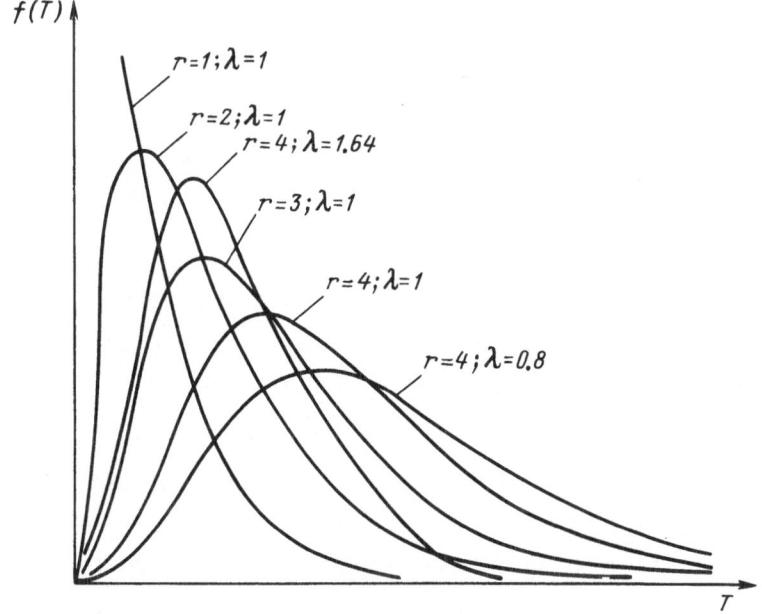

Fig. 27. The density of the gamma distribution for different values of r and λ.

The mathematical expectation $\underline{M}\{\tau\}$ and variance $\underline{D}\{\tau\}$ of the gamma distribution are equal to

$$\left.\begin{aligned} \underline{M}\{\tau\} &= \frac{r}{\lambda}, \\ \underline{D}\{\tau\} &= r/\lambda^2. \end{aligned}\right\} \tag{74}$$

To find $F(T)$ for integral values of r, we can use the nomogram shown in Fig. 28. The probability of failure-free operation in a period of time T

$$R(T) = \underline{P}\{\tau > T\} = 1 - F(T)$$

is plotted along the vertical axis. The quantity λT is plotted along the horizontal axis. Each curve of the nomogram represents a different value of r.

Example 3. From the values $\lambda = 0.1 \text{ h}^{-1}$, $r = 3$, find the probability of failure-free operation for $T = 8$ h.
By using the nomogram of Fig. 28, we find $\lambda T = 0.8$. From the point 0.8 on the horizontal axis, we draw a vertical line to its intersection with the line $r^* = r - 1 = 2$. Along the vertical axis we plot $R(T) = 0.95$. In the case of fractional r, we need to make a linear interpolation.

Let us now consider the question of estimating the quantities λ and r from experimental data. Depending on the form of the experimental data from which we estimate λ and r, we can use either of two approaches.
If it is possible to observe the sample functions of the wear of the type shown in Fig. 23 and if we know the value of M for the maximum admissible wear, then we can determine λ and r by making some simple calculations.
At the first stage of the calculation, we need to evaluate the mean rate of wear. We partition the entire interval of observation of the zone of normal wear (see Fig. 29) into several subintervals. The number of partition points is usually determined by the conditions of observation. As a rule, wear cannot be observed continuously. The quantities $\eta(t)$ for each of the objects vary over certain intervals of time. Therefore, natural intervals of observation are formed. Fig. 29 shows the results of measurement with the hollow dots. These hollow dots are connected by straight lines. It is desirable to measure the quantities $\eta(t)$ as frequently as possible. However, measuring difficulties make the intervals between measurements rather long. Furthermore, every measurement involves errors. If the measuring error is greater than the change in the quantity $\eta(t)$ since the preceding interval, this error can distort the curve representing the wear. Therefore, the interval between measurements must be such that the error in measuring will be small in comparison with the increase in the wear during that interval.

48

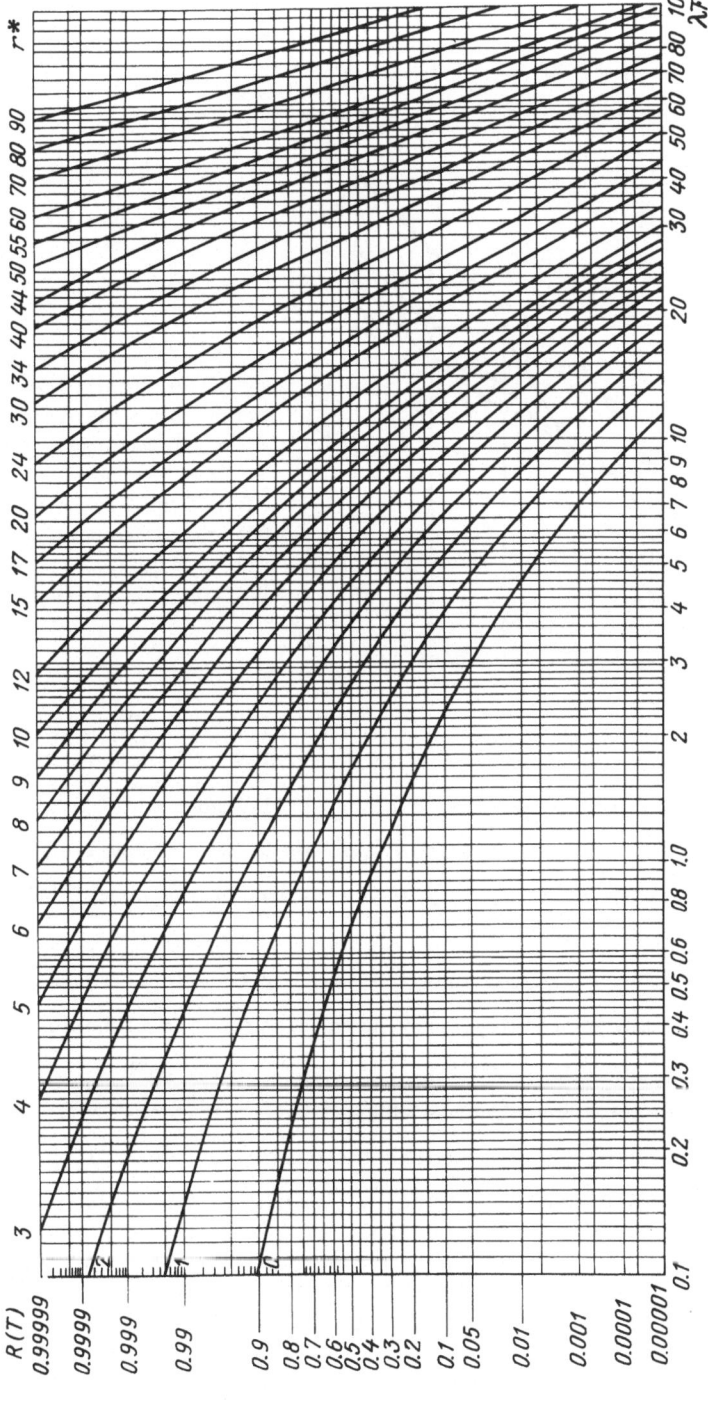

Fig. 28. A nomogram for determining the probability of failure-free operation $R(T) = P\{\tau > T\}$ from the gamma distribution; $r^* = r - 1$.

49

The total number of points of measurement must be at least 50.
Thus, if the number of objects studied is five, there must be at
least 10 measurements for each of them.

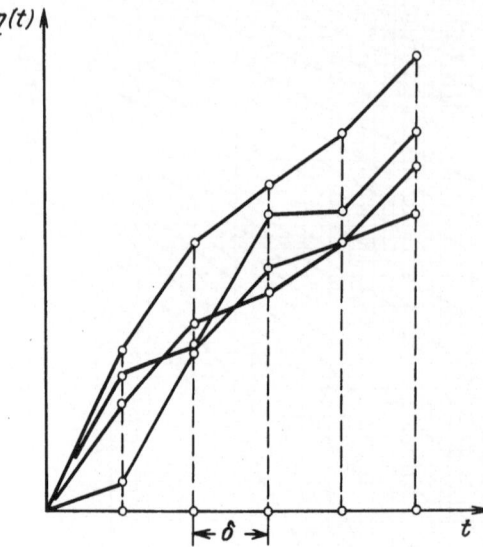

Fig. 29. The proce-
dure for determining
the parameters λ and
r from the sample
functions of the wear;
δ is the interval of
observation.

We confine ourselves to the case in which all the intervals of
observation have the same length δ. We shall denote by $\delta\eta_j^{(i)}$,
where the subscript j refers to the number of the interval of observation and the superscript i to the number of the object (the
sample function of the wear).
Up to now, we have assumed that the wear can be calculated as a
function of time. In practice, the wear can also occur as a function of
the number of cycles of loading, the length of the path (of a
flight), etc. Here, nothing has changed except that the probability of failure-free operation has the meaning of probability defined for a given number of cycles of loading or for a given
flight, etc.
Let us describe the order of analysis of the sample functions with
an example of data regarding the wear of tires (see Tab. 2). This
wear is determined by the radioactive method as a function of the
mileage.

Analysis of sample functions of wear. 1. For each of the objects, we calculate the mean value of increase in the wear $\overline{\delta\eta}^{(i)}$ during an interval of observation in accordance with the formula

$$\overline{\delta\eta}^{(i)} = \frac{1}{m} \sum_{j=1}^{m} \delta\eta_j^{(i)}, \tag{75}$$

where m is the number of intervals of observations.

2. We calculate the overall mean value of increase in the wear $\overline{\overline{\delta\eta}}$ in accordance with the formula

$$\overline{\overline{\delta\eta}} = \frac{1}{l} \sum_{i=1}^{l} \overline{\delta\eta}^{(i)}, \tag{76}$$

where l is the number of objects observed.

3. We calculate the variance s_i^2 of the quantity $\delta\eta_j^{(i)}$ and the mean variance in the increments $s_{\delta\eta}^2$ in accordance with the formulas

$$\left.\begin{array}{l} s_i^2 = \dfrac{1}{m-1} \sum_{j=1}^{m} (\delta\eta_j^{(i)} - \overline{\delta\eta}^{(i)})^2, \\[3mm] s_{\delta\eta}^2 = \dfrac{(m-1)\sum_{i=1}^{l} s_i^2}{ml-1} + \dfrac{m}{ml-1} \sum_{i=1}^{l} (\overline{\delta\eta}^{(i)} - \overline{\overline{\delta\eta}})^2. \end{array}\right\} \tag{77}$$

Before proceeding to a direct determination of the quantities λ and y, we need to show that the initial data do not contradict the assumption of homogeneity of the initial quality of the objects. If this is actually the case, there cannot be any significant difference between the quantities $s_1^2, s_2^2, \ldots, s_l^2$ or between the quantities $\overline{\delta\eta}^{(1)}, \overline{\delta\eta}^{(2)}, \ldots, \overline{\delta\eta}^{(1)}$. Mathematical statistics has available special criteria for testing the hypotheses of the equality of the variances and of the means. To verify the hypothesis that the variances are equal, we apply Bartlett's criterion [31], the procedure for using which is as follows:

We calculate the quantity

$$\chi^2 = \frac{2.3026}{1 + \dfrac{1+1}{31(m-1)}} \, l(m-1) \left[\log\frac{\sum_{i=1}^{l} s_i^2}{l} - \frac{1}{l} \sum_{i=1}^{l} \log s_i^2 \right] \tag{78}$$

and the so-called number of degrees of freedom

$$k = l - 1.$$

From a table of the χ^2 distribution (see [28], Tab. 4, p. 469), we find the value of χ_q^2 corresponding to the number of degrees of freedom k and the probability

$$\underline{P}\{\chi^2 > \chi_q^2\} = 0.05.$$

If it turns out that the value of χ^2 found in accordance with (78) does not exceed χ_q^2, that is, if $\chi^2 \leq \chi_q^2$ we assume that the experimental data do not contradict the hypothesis that the variances are equal.

The results of calculation of the quantities $\overline{\delta\eta}^{(i)}, \overline{\overline{\delta\eta}}, s_i^2$, and log s_i^2 for the sample functions of the wear of tires are shown in the last three columns of Tab. 2. With regard to this Table, $l = 5$ and $m = 10$. From formula (78), we obtain $\chi^2 = 3.2$ with $k = l - 1 = 4$ degrees of freedom.

From a table of values of χ^2, we obtain $\chi_{0.05}^2 = 9.5$, from which it follows that we have no reason for rejecting the hypothesis that the variances are equal.

4. The next step in statistical verification of the initial data is verification of the existence of divergent means. The procedure consists in the following.

We calculate the "sum of the squares of the deviations between series" Q_1:

$$Q_1 = m \sum_{i=1}^{l} (\overline{\delta\eta}^{(i)} - \overline{\overline{\delta\eta}})^2 \tag{79}$$

and the corresponding number of degrees of freedom $k_1 = l - 1$.

We calculate the "sum of the squares of the deviations within the series" Q_2:

$$Q_2 = (m - 1) \sum_{i=1}^{l} s_i^2 \tag{80}$$

and the corresponding number of degrees of freedom $k_2 = l(m - 1)$.

52

Table 2.   Increments in the wear of tires (mm) during an interval of observation $\delta = 1180$ km

Number of the object	Interval of observation of distance covered (km)										$\overline{\delta\eta}^{(i)}$	s_i^2	$\log s_i^2$
	0–1180	1180–2360	2360–3540	3540–4720	4720–5900	5900–7080	7080–8260	8260–9440	9440–10620	10620–11800			
i	$j=1$	$j=2$	$j=3$	$j=4$	$j=5$	$j=6$	$j=7$	$j=8$	$j=9$	$j=10$			
1	0.5	0.1	0.5	0.0	0.3	0.1	0.4	0.1	0.4	0.2	0.26	0.0338	$\overline{2}.5289$
2	0.6	0.5	0.2	0.1	0.3	0.5	0.2	0.1	0.6	0.4	0.35	0.0383	$\overline{2}.5832$
3	0.7	0.5	0.3	0.1	0.5	0.6	0.1	0.5	0.2	0.6	0.41	0.0477	$\overline{2}.6785$
4	0.6	0.7	0.0	0.3	0.3	0.2	0.7	0.0	0.0	0.2	0.30	0.0778	$\overline{2}.8910$
5	0.9	0.4	0.4	0.3	0.0	0.8	0.2	0.2	0.0	0.4	0.36	0.0893	$\overline{2}.9509$
											$\overline{\overline{\delta\eta}} = 0.336$	$\Sigma s_i^2 =$ $=0.2869$	$\Sigma \log s_i^2 =$ $=-6.3675$

$$s_{\delta\eta}^2 = 0.0554$$

The values $\overline{\delta\eta}^{(i)}, \overline{\overline{\delta\eta}}$, and s_1^2 were determined above by formulas (75)-(77).

We find the quantity

$$F = \frac{Q_1/k_1}{Q_2/k_2}. \tag{81}$$

From a table [4, Tab. 3.5, p. 270] of computed values of k_1 and k_2, we find the five percent point F of the distribution $F_{0.05}$. (In [4], the authors use v; instead of k_1.)

If the value of F found from [81] is less than $F_{0.05}(k_1;k_2)$, the hypothesis that the means are equal is taken; otherwise it is rejected.

Thus, from the data of Tab. 2, we have $Q_1 = 133 \cdot 10^{-3}, k_1 = 4$, $Q_2 = 2.58, k_2 = 45$. We find that $F = 0.58$. From the table [4], we have $F_{0.05}(4;45) \simeq 2.58$. Therefore, we take the hypothesis that the means are equal.

When we have convinced ourselves that the experimental data do not contradict the assumption that the original quality is the same for all the objects, we calculate λ and y from the formulas

$$\left. \begin{array}{l} y = \dfrac{s_{\delta\eta}^2}{\overline{\overline{\delta\eta}}}, \\[2em] \lambda = \dfrac{\overline{\overline{\delta\eta}}^2}{s_{\delta\eta}^2} \dfrac{1}{\delta}, \end{array} \right\} \tag{82}$$

where $s_{\delta\eta}^2$ and $\overline{\overline{\delta\eta}}$ are defined in accordance with (77) and (76). For the data of Tab. 2, we obtain $y = 0.165$ mm and $\lambda = 0.00173$ km^{-1}. Taking $M = 4$ mm in formula (70), we find $r = 24.2$.

Knowing r and λ, we can, as shown above, find the probability of failure-free operation over the given period of time.

Estimation of the parameters λ and r on the basis of data regarding the lifetime. Suppose that we have data on the lifetime of N objects $\tau_1, \tau_2, \ldots, \tau_N$.

1. We find the empirical mean and variance of the lifetime $\overline{\tau}$ and s_τ^2:

$$\left.\begin{array}{l} \overline{\tau} = \dfrac{1}{N} \sum\limits_{i=1}^{N} \tau_i, \\[4mm] s_\tau^2 = \dfrac{1}{N-1} \sum\limits_{i=1}^{N} (\tau_i - \overline{\tau})^2. \end{array}\right\} \qquad (83)$$

2. If we set $\overline{\tau} = \underline{M}\{\tau\}$ and $\underline{D}\{\tau\} = s_\tau^2$, we obtain relations from which we can find λ and r:

$$\left.\begin{array}{l} \lambda = \dfrac{\overline{\tau}}{s_\tau^2}, \\[6mm] r = \dfrac{\overline{\tau}^2}{s_\tau^2}. \end{array}\right\} \qquad (84)$$

Example 4. Data on the durability of cutting tools are shown in Tab. 3. The life length of a cutting tool was defined as the maximum admissible amount of wear. Find the values of λ and r and estimate the probability of failure-free operation for periods of time $T_1 = 60 \text{ min}$ and $T_2 = 100 \text{ min}$.

As a result of the calculations, we find $\overline{\tau} = 55.3$ and $s_\tau^2 = 620$.

From (84), we find $\lambda = 0.089$ and $r = 4.93$.

From the nomogram, we find $\underline{P}\{\tau > 60\} = 0.40$ and $\underline{P}\{\tau > 100\} = 0.06$.

Up to now, we have said nothing about the derivation of the gamma distribution. In the subsequent exposition, it will be useful to show the formal model leading to the arising of this distribution. Consider a system in which an accumulation of individual injuries takes place with the passing of time. If i injuries have occurred in the system up to the instant T, we shall say that it is in the state $\underline{E}_i$. The evolution of the system is described by the sequence

$$\underline{E}_0 \to \underline{E}_1 \to \underline{E}_2 \to \dots \to \underline{E}_r \to \dots \qquad (85)$$

Each individual injury occurs in accordance with a scheme of instantaneous damage. The probability of the transition $\underline{E}_k \to \underline{E}_{k+1}$ during the period Δ_T is given by the formula

$$\gamma(T) = \lambda \Delta_T + o(\Delta_T). \qquad (86)$$

Table 3. Data on the durability of cutting tools.

Number i of the object	τ_i (min)	Number i of the object	τ_i (min)
1	9	26	56.5
2	17.5	27	57.5
3	21	28	58
4	26.5	29	59
5	27.5	30	59
6	31	31	60
7	32.5	32	61
8	34	33	61.5
9	36	34	62
10	36.5	35	63
11	39	36	64.5
12	40	37	65
13	41	38	67.5
14	42.5	39	68.5
15	43	40	70
16	45	41	72.5
17	46	42	77.5
18	47.5	43	81
19	48	44	82.5
20	50	45	90
21	51	46	96
22	53.5	47	101.5
23	55	48	117.5
24	56	49	127.5
25	56	50	130

$$\overline{\tau} = 55.3; \quad s_\tau^2 = 620$$

The state of the system is characterized by the functions $\{\underline{P}_k(T), k = 0,1,\ldots\}$, where $\underline{P}_k(T)$ is the probability that the system will be in the state $\underline{E}_k$ by the instant T. We do not stop for a derivation of the expressions for $\underline{P}_k(T)$ (which can be found, for example, in [30], Chapter XVII) but present the final result:

$$\underline{P}_k(T) = \frac{(\lambda T)^k}{k!} e^{-\lambda T}, \quad k \geq 0. \tag{87}$$

The system of functions $\underline{P}_k(T)$ gives the distribution of an integer-valued random variable X(T) equal to the number of injuries sustained up to the instant T. Omitting the calculations, we present the formulas for the mathematical expectation and variance:

$$\underline{M}\{X(T)\} = \lambda T, \tag{88}$$

$$\underline{D}\{X(T)\} = \lambda T. \tag{89}$$

Remembering that every injury is the increment (representing the wear) in the variable y, we obtain the mean value of the wear during the period T:

$$\underline{M}\{X(T)y\} = y\lambda T \tag{90}$$

and the variance in the wear:

$$\underline{D}\{X(T)y\} = y^2 \lambda T. \tag{91}$$

These formulas were used to describe a way of estimating y and λ from the data on the wear; specifically, we obtained formulas (82) beginning with the equations

$$\left.\begin{array}{l} y\lambda\delta = \overline{\overline{\delta\eta}}, \\[2mm] y^2\lambda\delta = s^2_{\delta\eta}. \end{array}\right\} \tag{92}$$

Let us return to the question of the distribution of the lifetime τ. We recall that the lifetime is calculated up to the instant the rth injury (according to our numbering) is sustained. The sum of the probabilities $\underline{P}_0(T) + \underline{P}_1(T) + \ldots + \underline{P}_{r-1}(T)$ is the probability that the number of injuries sustained up to the instant T will be 0 or 1 ... or r - 1, that is, that it will be less than r. In other words, this is the probability that the lifetime τ will not be less than T; that is,

$$\underline{P}\{\tau > T\} = \sum_{k=0}^{r-1} \underline{P}_k(T). \tag{93}$$

From this we get, in accordance with formula (87),

$$\underline{P}\{\tau \le T\} = F(T) = 1 - \underline{P}\{\tau > T\} = 1 - \sum_{k=0}^{r-1} \frac{(\lambda T)^k}{k!} e^{-\lambda T}. \tag{94}$$

This is the desired formula. It coincides with (73) expressing the distribution function of the gamma distribution for integral r.

In the processing of experimental data, the number r may turn out
to be a fraction. For two reasons, this fact should not surprise
us. In the first place, the estimate of the number r on the basis
of experimental data is connected with random fluctuations and
therefore the deviation of the number r from an integer can be
random. In the second place, the number r usually proves to be
rather large (two-digit) and in practice we can always neglect
the error incurred in rounding it off. Therefore, we are justified
in considering the discrete model of the total number of injuries
suitable for describing a continuous scheme of actual wear.

We note that in the most general form, the density of the gamma
distribution is written as follows:

$$f(T) = \begin{cases} \dfrac{\beta^{\alpha+1}}{\Gamma(\alpha+1)} \, T^{\alpha} e^{-\beta T}, & T \geq 0, \\ 0 & , \ T < 0. \end{cases} \tag{95}$$

Here, α and β are the parameters of the distribution, arbitrary
positive numbers. By using tables (see [4], pp. 24, 202), we can
find the values of the integral function

$$\underline{P}\{\tau \leq u\} = \frac{1}{\Gamma(\alpha+1)} \int_{0}^{\beta u} y^{\alpha} e^{-y} \, dy \tag{96}$$

for given α, β, and u.

Normal distribution of lifetime

We noted above (see Fig. 27) that, with increasing r, the gamma
distribution becomes more symmetric. At the basis of this property
is the fact that, with increasing r, the curve for the density
$f(T)$ of the gamma distribution given by equation (67) tends to the
form

$$f(T) = \frac{1}{\sqrt{2\pi}\sqrt{r/\lambda^2}} \, \exp\left[- (T - r/\lambda)^2 / 2r\lambda^{-2} \right]. \tag{97}$$

In the general form, formula (97) can be represented as follows:

$$f(T) = \frac{1}{\sqrt{2\pi}\sigma} \exp\left[- (T - c)^2/2\sigma^2 \right], \qquad (98)$$

where c and σ are parameters.

One can easily see that the curve $f(T)$ in accordance with (98) is symmetric about the point $T = c$. Formula (98) gives the density of the normal distribution. A normal distribution is widespread in nature. Such random variables [10,28,31] as the errors in measurements, the errors in manufacturing, and various other errors have a normal distribution. Fig. 30 shows the density curves of a normal distribution. They are symmetric and have branches leading off in either direction to $T = \infty$ and $T = -\infty$. Thus, a normal distribution is defined on the entire T-axis from $-\infty$ to $+\infty$. At first glance, it seems strange that the normal distribution can describe the behavior of the lifetime τ since the time τ cannot be negative. Here, it is important to understand that the normal distribution yields an approximate (asymptotic) description of the distribution of the time τ. If we use a normal distribution for large values of r, we will always observe the fact that the probability of a negative value of τ, that is, the probability $\underline{P}\{\tau \leq 0\}$ given in the form

$$\underline{P}\{\tau \leq 0\} = \int_{-\infty}^{0} f(t)dt,$$

is an insignificantly small quantity and in practice will have no influence on the precision of the calculations of the reliability of objects. If the probability $\underline{P}\{\tau \leq 0\}$ proves to be a perceptibly large quantity, this means that we cannot use a normal distribution to describe the distribution of the time τ.

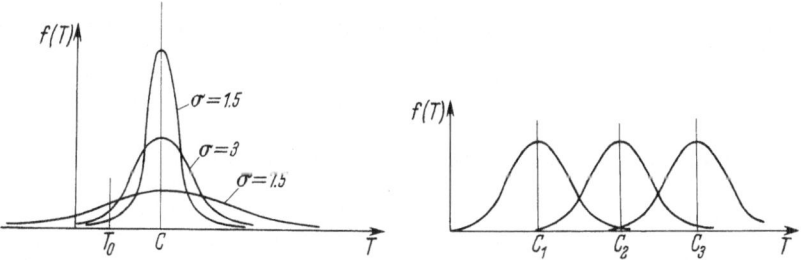

Fig. 30. Curves for the density of a normal distribution with different values of the parameters c and σ.

The mathematical expectation of a normal distribution having density (98) is equal to the parameter c, and its variance is equal to σ^2:

$$\left.\begin{array}{l} \underline{M}\{\tau\} = c, \\ \underline{D}\{\tau\} = \sigma^2. \end{array}\right\} \tag{99}$$

Combining (97) and (98), we obtain

$$\left.\begin{array}{l} c = \underline{M}\{\tau\} = \dfrac{r}{\lambda}, \\ \sigma^2 = \underline{D}\{\tau\} = \dfrac{r}{\lambda^2}. \end{array}\right\} \tag{100}$$

The normal distribution function can be calculated from the general rules to be

$$F(T) = \int_{-\infty}^{T} f(t)dt = \frac{1}{\sqrt{2\pi}} \int_{-\infty}^{(T-c)/\sigma} \exp(-u^2/2)du. \tag{101}$$

The function

$$\Phi(x) = \frac{1}{\sqrt{2\pi}} \int_{-\infty}^{x} \exp(-u^2/2)du \tag{102}$$

is called Laplace's function. Its values are given in the appendix (Tab. A1). Obviously,

$$F(T) = \Phi\left(\frac{T - c}{\sigma}\right). \tag{103}$$

Physically, the shift from the gamma distribution to a normal distribution is justified if the sample functions for the wear extend for a protracted period of time, weaving in and out of each other, before failures begin to occur. Formally, we can consider the shift to a normal distribution acceptable if the ratio

$$\frac{\underline{M}\{\tau\}}{\sqrt{\underline{D}\{\tau\}}} = \frac{r/\lambda}{\sqrt{r/\lambda^2}} > 3.5,$$

that is, if $r > 12$. The closeness of the gamma distribution to the normal distribution is illustrated by a comparison of the curves $R(T)$ (see Fig. 31). It is clear from the drawing that, even with

$r = 9$, the replacement of the gamma distribution with a normal distribution ensures maximum error in the estimate of $R(T)$ of the order of 10% almost everywhere except in the region of small probabilities.

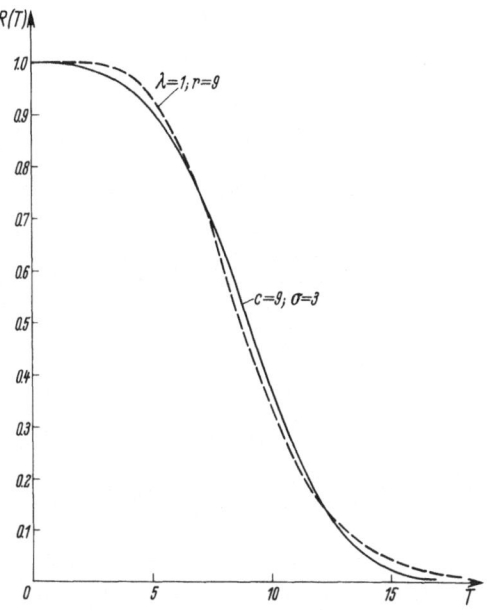

Fig. 31. Comparison of the curves for $R(T)$. The dashed curve represents a gamma distribution, the solid curve a normal distribution.

It is important to note that the normal distribution of the time τ comes about as a consequence of the homogeneity of the quality of the objects, the constant mean rate of wear, and the interweaving of the sample functions.

Let us point out an important characteristic of the normal distribution. For small values of $\sigma = \sqrt{D\{\tau\}}$ in comparison with the mean lifetime $c = \underline{M}\{\tau\}$, the values of the density are extremely close to zero on a rather long interval of time $(0, T_0)$ (see Fig. 30). It follows that, on a given interval, the probability of occurrence of a failure is very small. Physically, this reflects the fact that, if the maximum admissible wear M is great and the value of the total wear is small, then the probability of failure is small.

It is just this fact that is used for the introduction of compulsory replacements (repairs) at low wear levels, which ensures small probability of failure between the repairs. We note that, for an exponential distribution (see Fig. 3), which has maximum density at $T = 0$, the opposite is true; namely, a larger percent of the failures are observed during the initial period of operation.

The parameters c and σ of a normal distribution can be determined on the basis of experimental data regarding the wear or regarding the quantities τ_i. If we have the sample functions of the wear, then we need to find λ and r by the procedure described above and then determine c and σ from the formulas $c = r/\lambda$ and $\sigma^2 = r/\lambda^2$. If we have experimental data on the lifetime of N units $(\tau_1, \tau_2, \ldots, \tau_N)$, we need to calculate $\overline{\tau}$ and s_τ^2 and then take

$$c = \overline{\tau},$$
$$\sigma^2 = s_\tau^2.$$

Example 4 (continuation). From the data of Table 3, find the probabilities of failure-free operation during the periods $T_1 = 60$ min and $T_2 = 100$ min, assuming the distribution to be normal.
We set $c = \overline{\tau} = 55.3$ and $\sigma = s_\tau^2 = \sqrt{620} = 24.9$. We calculate

$$x_1 = \frac{T_1 - c}{\sigma} = 0.19; \quad x_2 = \frac{T_2 - c}{\sigma} = 1.8.$$

From Tab. A1, we find $\Phi(x_1) = 0.575$ and $\Phi(x_2) = 0.964$, from which we get $\underline{P}\{\tau > 60\} = 0.425$ and $\underline{P}\{\tau > 100\} = 0.036$.
From this it is clear that, for values of T close to the value of the mathematical expectation, the estimate for $\underline{P}\{\tau > T\}$ in accordance with a normal distribution is close to the estimate in accordance with the gamma distribution although r is small $(r = 4.93)$.

The shift from the gamma distribution to the normal distribution makes calculations much easier both because we have detailed tables of Laplace's function and because the functional form of this distribution is comparatively simple.

Sometimes, we assume that wear of an arbitrary form (any natural aging) leads to a normal distribution of the lifetime [1]. As one can see from the analysis that we have made, this distribution corresponds to a special form of the process of wear.

In conclusion, we point out two other ways of estimating the parameters c and σ of the normal distribution of the time τ. The first of these is associated with a partition of the set of values $\tau_1, \ldots, \tau_N$ into three subsets. We encountered a similar approach to the estimation of the parameters of the distribution in Chapter II when we effected a partition into two groups to estimate the parameter λ of the exponential distribution. The normal distribution has two parameters and this makes it necessary to partition the set of values τ_i into three subsets. In what follows, we shall refer to the method we have been describing for estimating the parameters of the distribution as the method of discrimination partitions.

<u>Estimation of the parameters c and σ of the normal distribution by the method of discrimination partitions.</u> Suppose that we have the data $\tau_1, \ldots, \tau_N$ on the lifetime of N units.

We choose numbers Θ_1 and Θ_2, where $\Theta_1 < \Theta_2$ and we calculate the number of units $m(\Theta_1)$ and $m(\Theta_2)$ that fail in the intervals $(0, \Theta_1)$ and $(0, \Theta_2)$.

We calculate the ratios

$$\nu(\Theta_i) = \frac{m(\Theta_i)}{N}, \quad i = 1, 2,$$

It is recommended that Θ_1 and Θ_2 be chosen in such a way that the inequalities $0.1 < \nu(\Theta_1) < 0.3$ and $0.5 < \nu(\Theta_2) < 0.9$ will be satisfied.

3. From Tab. A2 for the inverse Laplace function, we find the values $\Psi_1 = \Psi(\nu(\Theta_1))$ and $\Psi_2 = \Psi(\nu(\Theta_2))$.

4. The parameters c and σ are found from the formulas

$$c = \frac{\Theta_1 \Psi_2 - \Theta_2 \Psi_1}{\Psi_2 - \Psi_1}; \quad \sigma = \frac{\Theta_2 - \Theta_1}{\Psi_2 - \Psi_1}. \tag{104}$$

As we noted in Chapter II, the method of discrimination partitions is convenient when the estimate of the parameters of the distribution is made from the results of maintenance (testing) of the units of a given form in the process of use of the device. The quantity Θ_1 appears in this case as the time up to the first repair, and the quantity $\Theta_2 - \Theta_1$ appears as the time between the first and second repairs. Let n_1 and n_2 denote the numbers of units of a given form that are rejected in the first and second repairs, respectively. Let us assume that those devices in which defective units were discovered in the first test are not used anymore. A certain peculiarity in the calculation then arises. Specifically, in calculating the frequencies $\nu(\Theta_1)$, we need the formulas

$$
\left.
\begin{aligned}
\nu(\Theta_1) &= \frac{n_1}{N_1}, \\[2ex]
\nu(\Theta_2) &= \frac{n_1}{N_1} + \frac{n_2}{N_2}\left(1 - \frac{n_1}{N_1}\right),
\end{aligned}
\right\}
\tag{105}
$$

where N_1 is the number of units of a given form that are tested at the first repair and N_2 is the number of units of the given form tested at the second repair.

The smoothing of experimental data on the lifetime on probability paper for a normal distribution. We encountered probability paper in Chapter II in connection with the exponential distribu-

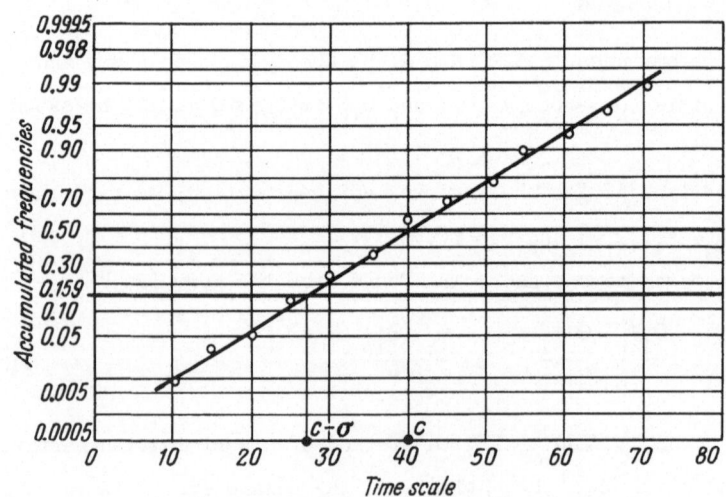

Fig. 32. Normal-distribution probability paper.

tion. To test for normality, we can smooth off the accumulated frequencies on probability paper for a normal distribution (see Fig. 32). If the experimental data agree sufficiently well with a normal distribution, the accumulated frequencies lie close to a straight line. On normal-distribution paper, we can estimate the parameters c and σ of the distribution. The parameter c is equal to the abscissa corresponding to the accumulated frequency 0.50, and the parameter σ is equal to the difference between the abscissas of points with accumulated frequencies 0.50 and 0.159. If the parameter r of the gamma distribution is small, this will be revealed when we do the smoothing off. Fig. 33 shows the results of smoothing off the gamma distribution on normal-distribution probability paper for the values $r = 4$ and $r = 8$.

The gamma distribution with bias parameter

Among the reasons leading to failures of devices and machines, one of the most important is the corrosion of metals. Corrosion takes place as a result of the action of the external medium on a metal and the contact between metals of different kinds. For a

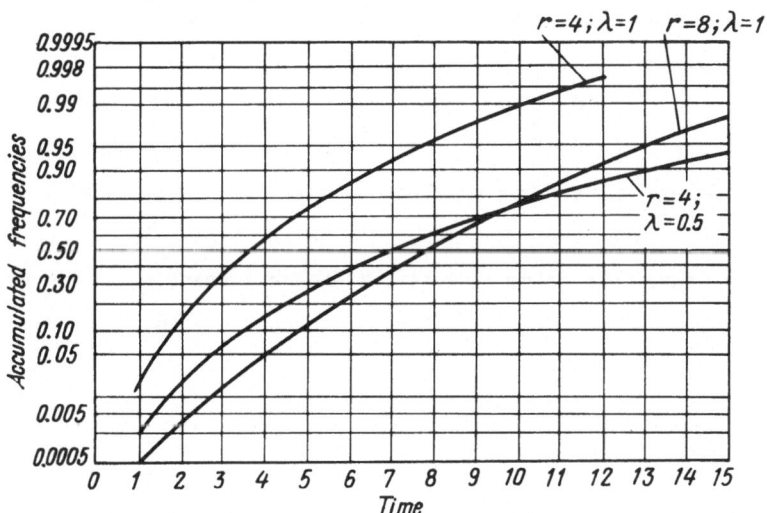

Fig. 33. Smoothing off of the gamma distribution on normal-distribution probability paper.

considerable portion of the forms of corrosion, the rate at which
it takes place is typically constant. Random variations in the
external conditions (changes in humidity, temperature, etc.) and
also the complexity of the internal structure of the metal cause
the rate of corrosion of each individual object to vary randomly
and causes the sample functions representing the progress of the
corrosion of objects of the same kind to become interwoven. Therefore, in many situations, failures that come about under the influence of corrosion correspond to conditions under which we have
a gamma distribution of a normal distribution.

To increase the reliability, we take special measures to protect
the objects from corrosion. These include protective coverings
(varnishes, paints, oxide coverings, etc.). Such a protective covering preserves the metal from corrosion for an extended period
of time. However, the covering itself is also subject to aging
and its protective properties gradually deteriorate. Consequently,
areas of corroded metal show up after the object has been in use
for a comparatively long period of time.

If we calculate the wear in terms of the weight loss of the metal
or in the depth of corrosion, then, when there is a protective
covering, we have a range where the value of the wear is equal to
zero (see Fig. 34a). The time t_0 until the beginning of corrosion
is random. Therefore, the set of sample functions of the wear will
have the form shown in Fig. 34b. Usually, the variation of the
time t_0 until the beginning of the wear is small in comparison
with its mean value. This enables us to assume that we are dealing
with a constant value t_0, which, just as in Chapter II, we shall
call the threshold of sensitivity. When the instant t_0 is reached,
wear begins. Of course, if failure can occur only as the result
of wear, then

$$\underline{P}\{\tau \le t_0\} = 0,$$

that is, the probability of failure prior to the instant t_0 is
zero.

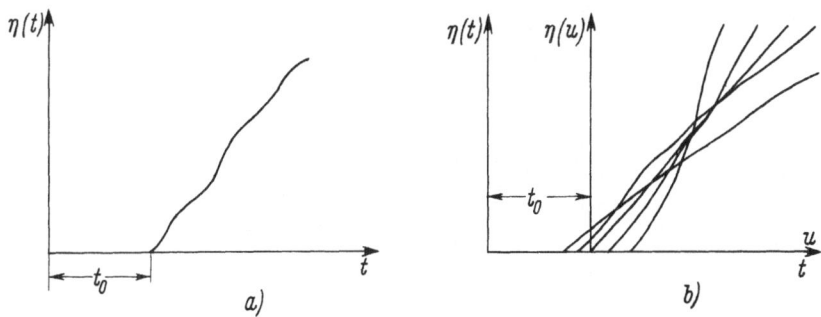

Fig. 34. A model for the occurrence of a gamma distribution with threshold of sensitivity.

If all the conditions for a gamma distribution, of which we spoke above, are satisfied then the probability density of the distribution of the time τ can be represented in the form

$$f(T) = \begin{cases} \dfrac{1}{\Gamma(r)} \lambda^r (T - t_0)^{r-1} e^{-\lambda(T-t_0)}, & T \geq t_0, \\ 0, & T < t_0. \end{cases} \tag{106}$$

If the threshold of sensitivity t_0 is given, then the parameters λ and r can be found by the procedures described above except that instead of t, we need to consider the time $u = t - t_0$. Thus, taking a new zero point for the time (see Fig. 34b), we obtain the probability density

$$f(u) = \begin{cases} \dfrac{1}{\Gamma(r)} \lambda^r u^{r-1} e^{-\lambda u}, & u \geq 0 \\ 0 & u < 0, \end{cases} \tag{107}$$

where, in analogy with (84),

$$\left. \begin{aligned} \lambda &= \frac{\overline{u}}{s_u^2}, \\ r &= \frac{s_u^2}{\overline{u}^2}. \end{aligned} \right\} \tag{108}$$

When we are given sample functions of the wear, the quantity t_0 can be determined approximately by constructing the mean curve of wear (see Fig. 35). The mean curve of wear can be drawn by the method of least squares [22]. The point of intersection of the mean curve of wear with the $\underline{t}$-axis gives the value of the threshold of sensitivity t_0.

When we have data only concerning the quantities $\tau_1,\ldots,\tau_N$, that is, the lifetimes of N objects, the determination of the threshold t_0 involves considerable difficulties.

The method of finding the quantities t_0, λ, and r is based on equating the theoretical and empirical central moments of the first three orders. (This is called the method of moments.) We note that, by virtue of the displacement $\tau = U + t_0$,

$$
\left.
\begin{aligned}
\underline{M}\{\tau\} &= \underline{M}\{u\} + t_0 = \frac{r}{\lambda} + t_0, \\[2mm]
\underline{D}\{\tau\} &= \underline{D}\{u\} = \frac{r}{\lambda^2}.
\end{aligned}
\right\}
\tag{109}
$$

By using formula (8), we can show that the third-order central moment is equal to

$$
\mu_3 = \frac{2r}{\lambda^2}.
\tag{110}
$$

In the Introduction, we pointed out that its empirical analogue m_3 is defined by the expression (15). When we equate the values of $\overline{\tau}$, s_τ^2, and m_3 found from the experimental data with the corresponding theoretical values of the moments, we get the following formulas after some simple transformations:

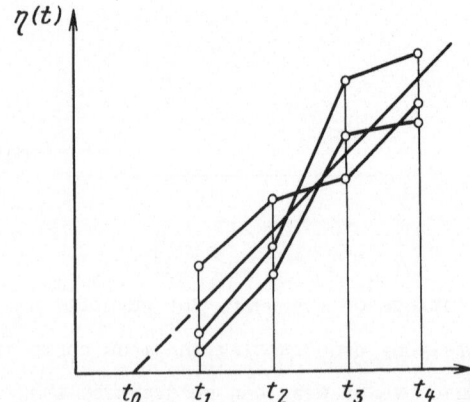

Fig. 35. Estimation of the threshold of sensitivity from sample functions.

$$\lambda = 2\frac{s_\tau^2}{m_3},$$

$$r = 4\frac{[s_\tau^2]^3}{(m_3)^2},$$

$$t_0 = \overline{\tau} - \frac{2(s_\tau^2)^2}{m_3}.$$

(111)

The estimate m_3 of the third-order central moment has great random variations. Therefore, this estimate is rather crude in the statistical sense. It can lead to large random variations in the estimates of the quantities t_0, λ, and r. Therefore, when we process the experimental data in accordance with formulas (111), we can come to the conclusion that the threshold of sensitivity $t_0 > 0$ although in actuality it is equal to zero. By virtue of this, when we take into account the threshold of sensitivity in processing experimental data, we need first of all to rest on the physical picture of a failure. Furthermore, it is necessary to have at our disposal at least 100 items of information regarding the variables τ_i. Finally, we always need to take $t_0 = 0$ if the results of processing the experimental data indicate that

$$t_0 < 0.1\,\overline{\tau}.$$

When we have a large number of items regarding the quantities τ_i, we can obtain a representation regarding the existence and value

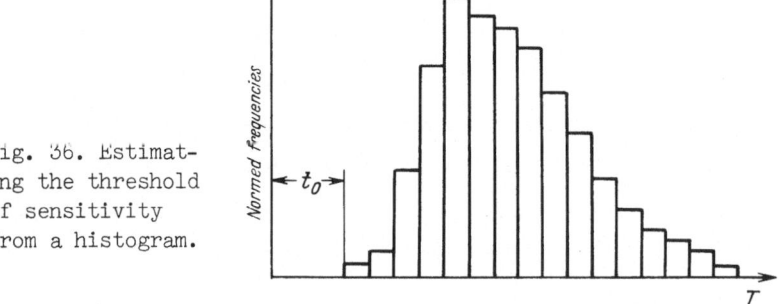

Fig. 36. Estimating the threshold of sensitivity from a histogram.

69

of the threshold of sensitivity by constructing a histogram of the
distribution (see Fig. 36). The characteristic displacement of the
entire histogram to the right from its position at $t_0 = 0$ should be
noted.

As we have already mentioned, for large values of r, the gamma
distribution can be replaced with a normal distribution. This property remains true when there is a threshold of sensitivity. If
the gamma distribution is close to a normal distribution, there is
no need to introduce the threshold t_0. Here, just as before, we
use the formulas $c = \bar{\tau}$ and $\sigma^2 = s_\tau^2$ to estimate the parameters of the
normal distribution. Consequently, it is worthwhile to resort to
a determination of the threshold t_0 from the experimental data regarding the quantities τ_i and to estimate λ and r from formulas
(111) only when the histogram of the distribution has an asymmetric form and is displaced to the right. To avoid error, it is
expedient to plot these data on normal-distribution paper whenever
one is smoothing out experimental data with the aid of the gamma
distribution. If the accumulated frequencies lie close to a
straight line, it makes sense to shift from the gamma distribution
to a normal distribution.

The functional gamma distribution

In the preceding schemes for a process of wear, the assumption
that the mean rate of wear is constant was essential. However, as
we have pointed out, such constancy is by no means always the case.
In particular, the mean rate does not remain constant when corrosion is taking place.

We know, that a number of metals, for example, aluminum and its
alloys have a hard and relatively dense oxide. When pure aluminum
is subjected to oxidation, the surface becomes covered with a layer of aluminum oxide Al_2O_3, which has a high degree of hardness
and density. This layer of oxide, formed under the action of the
atmosphere, hinders further penetration of oxygen and thus hinders
further corrosion. The denser the layer of the oxide, the slower

will subsequent corrosion take place. This leads to the fact that the rate of corrosion decreases (see [29], p. 51) as

$$\frac{d\underline{M}\{\eta(t)\}}{dt} = \frac{a}{b+t} , \qquad (112)$$

where a and b are constants and t is the time the object is in use (see Fig. 37).

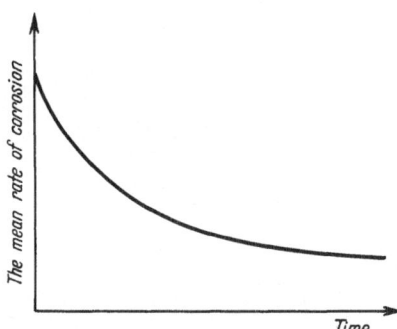

Fig. 37. The behavior of the mean rate of corrosion.

A process of corrosion is not the only example of decrease in the mean rate of wear. A number of such processes possess this property. These are essentially diffusion processes of some sort or other. Examples are processes of accumulation of fatigue in metals, processes of accumulation of deformation at high temperatures (creep, extended durability [stress-rupture]) and a number of others.

It is not difficult to imagine a situation in which the rate of wear increases with the passage of time. This is the case when other processes, for example, increase in temperature, vibration, etc., are superimposed on the wear-process.

Let us suppose that we have at our disposal a pencil of sample functions of some process of wear (see Fig. 38), the mean rate of which is a known function of the time v(t):

$$v(t) = \underline{M}\{\xi(t)\}. \qquad (113)$$

The increase in the wear is, on the average, described by the function u(t):

$$\underline{M}\{\eta(t)\} = u(t) = \int_0^t v(z)dz, \qquad (114)$$

where $u(t)$ is assumed to be a monotonic function of t.

Let us assume that all the sample functions correspond to objects having the same initial quality. This means that the sample functions representing the rate of wear $\xi_i(t)$ (where i is the number

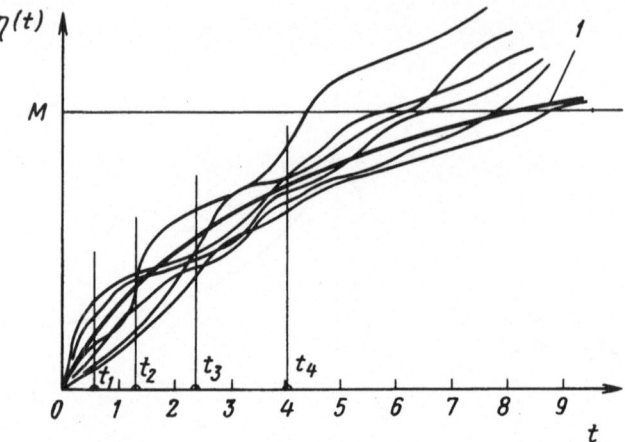

Fig. 38. Sample functions of the wear with variable mean rate. The curve 1 is the curve of mean accumulated wear.

of the sample function) are sample functions of the same random process and that they differ from each other only in the random sense. Let us suppose that, for the case described, the relation (51) holds, that is, that

$$\xi(t) = v(t)\rho(t), \qquad (115)$$

where $\rho(t)$ is a stationary random process, and $\underline{M}\{\rho(t)\} = 1$.

The first factor reflects the determined change in the rate on the average and the second represents the random variations about this mean level.

Let us suppose that the process $\rho(t)$ is a process with decreasing connections. Physically, this means that the behavior of $\rho(t)$ during the interval $(T_2, T_2 + \Delta_T)$ depends very weakly on its behavior in the interval $(T_1, T_1 + \Delta_T)$. Let us clarify the physical meaning of this assumption with an example of a corrosion process.

Corrosion is caused by the interaction between a metal and the ex-

ternal medium. The variations in its rate are to a considerable
degree determined by the variations in the properties of the external medium. Parameters of the external medium such as moisture,
saturation with chemically active materials, temperature, etc.,
change continuously and randomly. It is just these changes that
cause a "superposition" of the random component $\rho(t)$ on the mean
rate of the corrosion process. However, random properties of the
external medium during the intervals $(T_1, T_1 + \Delta_T)$ and $(T_2, T_2 + \Delta_T)$
are connected only very weakly if $(T_2 - T_1)$ is sufficiently great.
This leads to the fact that the process $\rho(t)$ has decreasing connections.

Let us now make a change in the time; specifically, let us introduce a new variable time $u = u(t)$ and let us construct our pencil
of sample functions in the new time. This means that, when we construct the abscissa $u(t_1)$, we make it correspond to the ordinate
$\eta(t_1)$. Obviously, the sample functions of the wear will now have
a constant mean rate. The properties of independence (or quasi-
independence) of the increments in the wear $\eta(t)$ and the homogeneity of the initial quality are maintained. Consequently, everything
that was said in the derivation of the gamma distribution remains
applicable for such sample functions. In particular, the random
time of attainment of a level M has a gamma distribution with density

$$
f_1(u) = \begin{cases} \dfrac{1}{\Gamma(r)} \, \lambda^r u^{r-1} e^{-\lambda u}, & u \geq 0, \\ 0 & , \ u < 0. \end{cases}
\tag{116}
$$

Let us show what the distribution of the "true" lifetime will be
like. We are dealing with a situation in which we know the distribution density of the quantity u and we need to find the density
of the quantity τ, where $u = u(\tau)$ is a known monotonic function.
We have

$$
\underline{P}\{\tau \leq T\} = \underline{P}\{u \leq u(T)\} = \underline{P}\left\{u \leq \int_0^T v(t)dt\right\}.
\tag{117}
$$

The expression (116) enables us to determine the distribution function of the random variable u:

$$P\{u \leq U\} = F(U) = \int_0^U f_1(u)du. \tag{118}$$

By comparing (117) and (118), we obtain a relation for the distribution function $G(T)$ of the random variable τ:

$$G(T) = P\{\tau \leq T\} = F(\int_0^T v(t)dt) = F(u(T)). \tag{119}$$

Let us look at the case (important in applications) when

$$M\{\xi(t)\} = \frac{a}{1+t}, \quad a > 0. \tag{120}$$

For this case, we have, in accordance with (114),

$$u(t) = a \ln(1+t). \tag{121}$$

In accordance with (119),

$$G(T) = F(a \ln(1+T)) = \int_0^{a \ln(1+\tau)} f_1(u)du. \tag{122}$$

From this we easily obtain by differentiation the distribution density $g(T)$ of the random variable τ:

$$g(T) = \begin{cases} \dfrac{1}{\Gamma(r)} \lambda^r a^r [\ln(1+T)]^{r-1} e^{-\lambda a \ln(1+T)} \dfrac{1}{1+T}, & T \geq 0, \\ 0, & T < 0. \end{cases} \tag{123}$$

The graph of the density of this distribution is sharply asymmetric and is pulled to the side of large values of T (see Fig. 39). We arrive at the same distribution if we assume that, during the time from T to $T + \Delta_T$, the object sustains a single injury with probability

$$\gamma(T) = \frac{a\lambda}{1+T} \Delta_T + o(\Delta_T), \tag{124}$$

and that failure occurs when a total of r injuries has been sustained. An analogous situation was described on page 56. For the present case, we can also derive the formulas for $P_k(T)$, that is, the probability that exactly k injuries will have occurred up to

74

the instant T. These formulas differ from formula (87) in that instead of λt we have the quantity

$$\int_0^T \frac{a\lambda}{1+t}\, dt = \lambda a \ln(1+T):$$

Specifically,

$$P_k(T) = \frac{\left[a\lambda \ln(1+T)\right]^k}{k!}\, e^{-a\lambda \ln(1+T)}, \quad k \geq 0. \qquad (125)$$

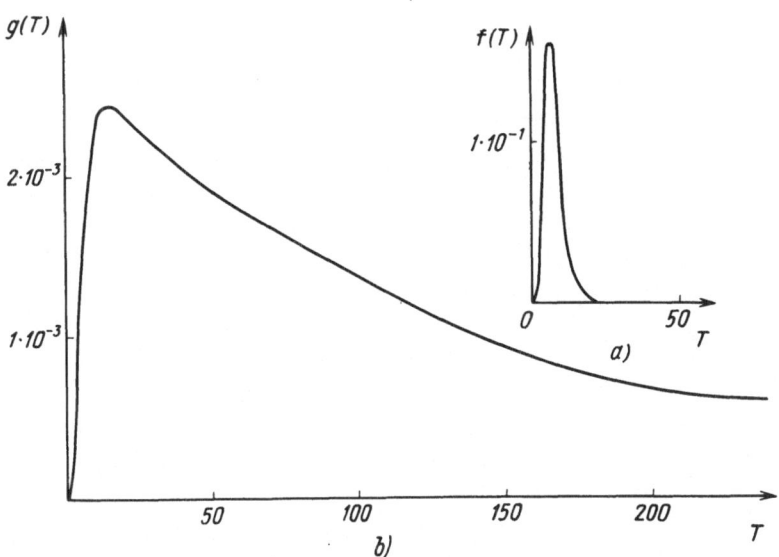

Fig. 39. Comparison of the curves of the density of the gamma distribution (a) and the functional gamma distribution (b) for $a = \lambda = 1$, $r = 7$.

In accordance with formula (94), we get an expression for the distribution function of τ:

$$P\{\tau \leq T\} = G(T) = 1 - \sum_{k=0}^{r-1} P_k(T). \qquad (126)$$

From this we obtain by differentiation a formula for the density coinciding with the expression (123).

Features of the processing of experimental data. If we have sample functions of the wear, then, to determine the quantities y, λ, and r, we need to use the procedure expounded above. In this case, the only peculiarity of the processing of the data consists in the

fact that the sample functions of the wear should be constructed
in a time scale defined by

$$u(t) = \int_0^t v(z)dz.$$

Thus, if the mean rate of wear is given by formula (120), when we
construct the sample functions of the wear on the abscissa we need
to plot not the time of observation t itself but the quantity <u>a</u>
ln (1 + t). Here, the wear has on the average a constant rate. Ac-
cordingly, formulas (82) remain valid without modification pro-
vided the observations on the objects are carried out over intervals of time Δ that are constant in length in the time scale u.
This can be done if the measurements of the wear are carried out
at instants of time t_i chosen in such a way that the difference
ln $(1 + t_{i+1})$ - ln $(1 + t_i)$ is constant. This means that $t + 1$ points
of measurement of the wear form a geometric progression on the
scale (see Fig. 38).

If the experimental data consists of the values $\tau_1, \tau_2, \ldots, \tau_N$ of
the lifetime of N exemplaires, then, when we are processing this
data, we need first to shift from the values τ_i to the quantities

$$u_i = \int_0^{\tau_i} v(z)dz,$$

then find the mean and empirical variance $\bar{u}$ and s_u^2, and finally
calculate λ and r in accordance with formulas (84). As can be seen
from what has been said, the nonconstancy of the mean rate of wear
$\underline{M}\{\xi(T)\}$ does not cause particular complications in processing the
experimental data.

The logarithmic normal distribution

We spoke above of the fact that, for large values of r, the gamma
distribution approximates a normal distribution. This suggests
that we might repeat the calculations of the preceding section,
assuming that the random variable u has a density of the form

$$f(u) = \frac{1}{\sqrt{2\pi}\sigma} \exp[-(u-c)^2/2\sigma^2]. \tag{127}$$

Suppose that the mean rate of wear is given by the formula

$$\underline{M}\{\xi(t)\} = \frac{a}{1+t}, \quad a > 0. \tag{128}$$

If we repeat step by step the reasoning of the preceding section, we obtain a formula analogous to formula (122):

$$\underline{P}\{\tau \leq T\} = G(T) =$$

$$= \frac{1}{\sqrt{2\pi}\sigma} \int_{-\infty}^{a \ln(1+T)} \exp[-(u-c)^2/2\sigma^2] du = \Phi\left[\frac{a \ln(1+T) - c}{\sigma}\right], \tag{129}$$

where $\Phi(x)$ is Laplace's function defined by formula (102). From this we get the distribution density

$$g(T) = \begin{cases} \frac{1}{\sqrt{2\pi}\sigma} \exp\{-[a \ln(1+T) - c]^2/2\sigma^2\} \frac{a}{1+T}, & T \geq -1 \\ 0 & , T < -1. \end{cases} \tag{130}$$

The distribution that we have obtained is called the logarithmic normal distribution. We must not be confused by the fact that $g(T)$ is positive on the interval $(-1, \infty)$ rather than merely on the interval $(0, \infty)$ as has been the case up to now. This is due to the fact that we are using the density of the normal distribution (127) instead of the gamma distribution since the density of the normal distribution is defined on the interval $(-\infty, +\infty)$. This fact is not significant since, as a rule, $c \gg 1$ and therefore the "tail" of $g(T)$ is not perceptible in the negative region. It is customary to write the density of the logarithmic normal distribution in the form

$$g(T) = \begin{cases} \frac{A}{\sqrt{2\pi}\sigma} \exp[-(\log T - c)^2/2\sigma^2]\frac{1}{T}, & T \geq 0, \\ 0, & T < 0, \end{cases} \tag{131}$$

where

$$A = \log e = 0.4343.$$

This formula is the same as formula (130) except that we replace

1 + T with T and the natural with the common logarithm and we set a = 1.

The curves of the density of the logarithmic-normal distribution are shown in Fig. 40. These curves are asymmetric; the vertex of the density curve lies to the left of the mathematical expectation. The asymmetry becomes the more noticeable when the value of the parameter σ is large.

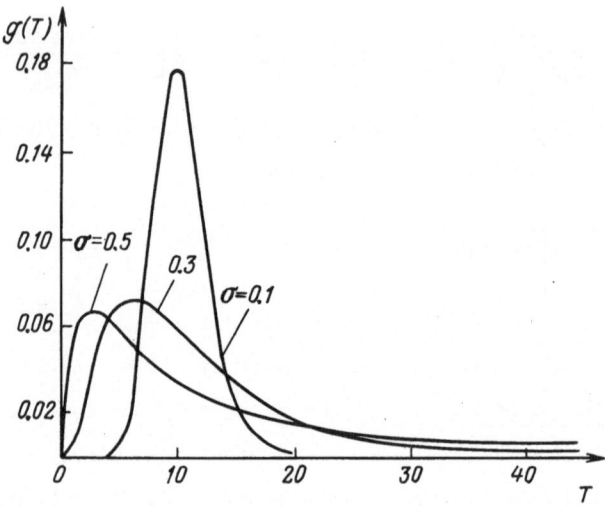

Fig. 40. The curves for the density of the logarithmic normal distribution with c = 1.

The logarithmic normal distribution is widely used in reliability theory. It is used for the processing of experimental data regarding the fatigue longevity of metals, the extended durability, the lifetime of a number of units in radio-electronic equipment, and other cases.

In what follows, in processing experimental data we shall use the density g(T) in the form given by formula (131).

Let us pause to look at the properties of the logarithmic normal distribution. The parameters c and σ of the distribution (131) are connected with the mathematical expectation and variance of the random variable τ as follows:

$$\underline{M}\{\tau\} = \exp\left[\frac{c}{A} + \frac{\sigma^2}{2A^2}\right],$$

$$\underline{D}\{\tau\} = \exp\left[2\frac{c}{A} + \frac{\sigma^2}{A^2}\right]\left[\exp\left(\frac{\sigma^2}{A^2}\right) - 1\right], \qquad (132)$$

where $A = 0.4343$.

The mode (maximum of the density curve) has abscissa t_M, the logarithm of which equal to

$$\log t_M = c - 2.3026\sigma^2. \qquad (133)$$

If we transform the quantity τ with density (131) into the quantity $u = \log \tau$, this new quantity has a normal distribution with parameters $\underline{M}\{u\} = c$ and $\underline{D}\{u\} = \sigma^2$. This last property is especially important and is widely used in the processing of experimental data.

Estimation of the parameters of the logarithmic normal distribution from the data on the lifetime. If the experimental data consists in the quantities $\tau_1, \ldots, \tau_N$ representing the lifetimes of N objects, we proceed as follows to estimate the parameters c and σ.

1. From the values of τ_i, we find the quantities $u_i = \log \tau_i$, and then we find their means $\bar{u}$ and their variances s_u^2 from the formulas

$$\bar{u} = \frac{\sum\limits_{i=1}^{N} u_i}{N},$$

$$s_u^2 = \frac{1}{N-1}\sum\limits_{i=1}^{N}(u_i - \bar{u})^2. \qquad (134)$$

2. We obtain the parameters c and σ from the equations

$$\begin{aligned} c &= \bar{u}, \\ \sigma^2 &= s_u^2, \end{aligned} \qquad (135)$$

which follow from the fact that $u = \log \tau$ has a normal distribution. Furthermore, we can use the method of discrimination partitions or smoothing on normal paper as described above. Both these methods must be applied to the variables u_i, where we always have

$$\left.\begin{array}{l} \underline{M}\{u\} = c, \\ \underline{D}\{u\} = \sigma^2. \end{array}\right\} \tag{136}$$

The probability of failure-free operation in the course of the time T is calculated in accordance with the formula

$$\underline{P}\{\tau > T\} = R(T) = 1 - \Phi\left(\frac{\log T - c}{\sigma}\right). \tag{137}$$

Example 5. Objects are tested under a stress $\sigma_{max} = 30$ kg/mm^2. The distribution is the longevity distribution (131). Find the parameters c and σ.

Table 4. Results of tests for fatigue (brand alloy B-95, longitudinal direction, cantilever bending, $\sigma_{-1} = 20$ kg/mm^2) [27, p. 109]

$\sigma_{max} = 21$ kg/mm^2			$\sigma_{max} = 30$ kg/mm^2		
Number of cycles prior to a failure $N_i \cdot 10^{-5}$	Accumulated frequencies $\nu(N_i)$	$\log N_i = u_i$	Number of cycles prior to a failure $N_i \cdot 10^{-5}$	Accumulated frequencies $\nu(N_i)$	$\log N_i = u_i$
6.64	0.023	5.822	0.53	0.023	4.724
7.13	0.068	5.853	0.65	0.068	4.813
7.88	0.114	5.896	0.76	0.114	4.881
7.97	0.159	5.902	0.80	0.159	4.905
7.99	0.204	5.903	0.87	0.204	4.940
10.65	0.250	6.027	0.90	0.250	4.954
12.08	0.295	6.082	0.90	0.295	4.954
12.53	0.341	6.098	1.02	0.341	5.009
13.08	0.386	6.117	1.07	0.386	5.029
15.90	0.432	6.201	1.07	0.432	5.029
17.08	0.477	6.232	1.09	0.477	5.037
17.62	0.523	6.246	1.16	0.523	5.064
24.90	0.568	6.396	1.22	0.568	5.086
30.20	0.614	6.480	1.29	0.614	5.111
38.34	0.659	6.584	1.40	0.659	5.146
50.09	0.704	6.700	1.57	0.704	5.196
59.90	0.750	6.777	1.59	0.750	5.201
78.32	0.795	6.894	1.88	0.795	5.274
97.13	0.841	6.987	2.07	0.841	5.316
176.20	0.886	7.246	2.23	0.886	5.348
327.01	0.932	7.515	2.38	0.932	5.377
346.84	0.977	7.540	2.79	0.977	5.446

$$\bar{u} = 5.084$$
$$s_u = 0.189$$

From Tab. 4, we find the values of the logarithms of the longevity $\log N_i$ and the accumulated frequencies $\nu(N_i)$. Fig. 41 shows that they can be smoothed sufficiently well on normal paper. The distinctive features of the smoothing in this case are that the quantities $u_i = \log N_i$ are laid off along the abscissa. From Fig. 41, we obtain the values $c = 5.08$ and $\sigma = 0.17$.
Let us use the method of discrimination partitions. We choose as points of the partition $u^*_1 = 5.0$ and $u^*_2 = 5.3$. Since 7 and 18 values of the logarithms of the longevity lie to the left of u^*_1 and u^*_2, respectively, we have

$$\nu(u^*_1) = \frac{7}{22} = 0.32; \quad \nu(u^*_2) = \frac{18}{22} = 0.82.$$

From Tab. A2, we find

$$\Psi_1 = \Psi(0.32) = -0.468; \quad \Psi_2 = \Psi(0.82) = 0.915.$$

From formulas (104), we find $c = 5.10$ and $\sigma = 0.217$.
Let us compare the results obtained with the values of c and σ calculated directly from formulas (136). The values of $\overline{u}$ and s_u are given in Tab. 4. We have $c = 5.084$ and $\sigma = 0.189$.

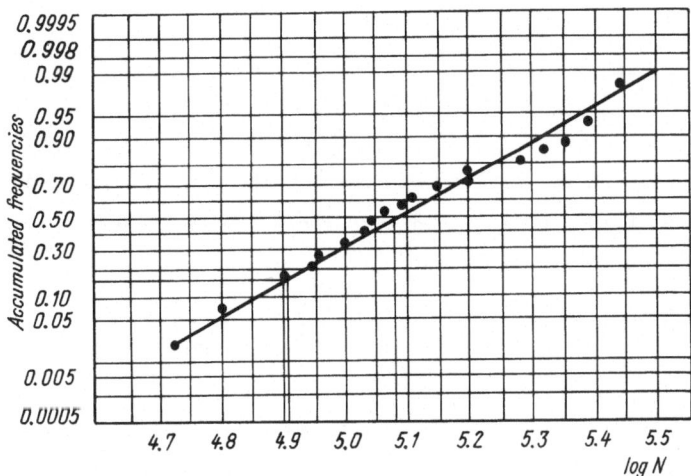

Fig. 41. Smoothing of the data on the longevity of objects under heavy loads on normal-distribution probability paper.

We need to keep in mind the fact that variations in the estimates of c and σ are unavoidable. The most precise estimates are obtained from formulas (136). It must be acknowledged that the estimate of the parameters with the use of probability paper yields an acceptable degree of accuracy.

In a number of cases, the process of accumulation of injuries (wear and tear) begins only after some instant t_0 subsequent to the beginning of use. The quantity t_0 is called the threshold of sensitivity. We have spoken of it already in connection with the gamma distribution, which has a displacement parameter. The existence of a threshold of sensivity is typical, for example, for a process of accumulation of fatigue injuries under low stresses. This is caused by the existence of a period of incubation during which the method receives no fatigue injuries.

It follows from what we have said that the mathematical expactation of the rate of wear is given by the equation

$$\underline{M}\{\xi(t)\} = \begin{cases} 0 & , \ t < t_0, \\ \dfrac{a}{1 + t - t_0}, & t \ge t_0. \end{cases} \tag{138}$$

Now, $u(T)$ can be written in the form

$$u(T) = \begin{cases} 0 & , \ T < t_0, \\ a \ln(1 + T - t_0), & T \ge t_0. \end{cases} \tag{139}$$

When we substitute this expression into the upper limit of integration in (129), we arrive at a formula for the density

$$g(T) = \begin{cases} \dfrac{1}{\sqrt{2\pi}\sigma} \exp\left\{ -\dfrac{[a \ln(1 + T - t_0) - c]^2}{2\sigma^2} \right\} \dfrac{a}{1 + T - t_0}, & T \ge t_0 - 1, \\ 0, & T < t_0 - 1. \end{cases} \tag{140}$$

This formula can be simplified if we shift from T to the variable $T' = T + 1$ and from natural to common logarithms. For simplicity, we keep the same letter for the argument of the function. Then,

$$g(T) = \begin{cases} \dfrac{\log e}{\sqrt{2\pi}\,\sigma} \exp\left\{ -\dfrac{[\log(T - t_0) - c]^2}{2\sigma^2} \right\} \dfrac{1}{T - t_0}, & T \ge t_0, \\ 0, & T < t_0. \end{cases} \tag{141}$$

82

This is the generally used representation of the logarithmic normal distribution with displacement parameter t_0 (the threshold of sensitivity).

The smoothing of experimental data and estimation of the parameters from data on the lifetime. Suppose that we have data on the lifetime of N objects $\tau_1,\ldots,\tau_N$. The parameters, c, σ, and t_0 can be estimated in the following way [1].

1. We choose a number q in the interval $0 < q < 1/2$. It is suggested that one choose q about equal to 0.1 or 0.2.

2. From Tab. A2, we find the value $\Psi(q)$ of the inverse Laplace function.

3. We arrange all the values τ_i in increasing order:

$$\tau_1 \leq \tau_2 \leq \tau_3 \leq \ldots \leq \tau_N,$$

and calculate the accumulated frequencies $\nu(\tau_i)$ (see Tab. 1 and the explanation on p. 23).

4. We find empirical quantiles $\tau_q, \tau_{1/2}$, and τ_{1-q} corresponding to to levels q, 1/2, and 1-q. This is done as follows: For the number q, we find the accumulated frequencies $\nu(\tau_i)$ and $\nu(\tau_{i+1})$ such that q will lie between them:

$$\nu(\tau_i) \leq q \leq \nu(\tau_{i+1}).$$

For τ_q we find the number

$$\tau_q = \tau_i + \frac{\tau_{i+1} - \tau_i}{\nu(\tau_{i+1}) - \nu(\tau_i)} (q - \nu(\tau_i)). \tag{142}$$

We proceed analogously with the numbers 1/2 and 1 - q. [2]

---

[1] Below, we shall expound the method of quantiles, which is the simplest for calculations. This method is described in [36, p. 58].

[2] The quantile $\tau_{1/2}$ is called median.

5. We calculate σ, c, and t_0 from the formulas

$$\left.\begin{array}{l} \sigma = \dfrac{\log x}{\Psi(q)}, \\[3mm] c = \log \dfrac{\tau_{1/2} - \tau_q}{1 - x}, \\[3mm] t_0 = \tau_{1/2} - \dfrac{\tau_q - \tau_{1/2}}{x - 1}, \end{array}\right\} \qquad (143)$$

where

$$x = \frac{\tau_q - \tau_{1/2}}{\tau_{1/2} - \tau_{1-q}}. \qquad (144)$$

The mathematical expectation and variance of the time τ is expressed in terms of the parameters c, σ, and t_0:

$$\left.\begin{array}{l} \underline{M}\{\tau\} = t_0 + \exp\left(\dfrac{c}{A} + \dfrac{\sigma^2}{2A^2}\right), \\[5mm] \underline{D}\{\tau\} = \exp\left[2\dfrac{c}{A} + \dfrac{\sigma^2}{A^2}\right]\left[\exp\left(\dfrac{\sigma^2}{A^2}\right) - 1\right] \end{array}\right\} \qquad (145)$$

where $A = \log e = 0.4343$.

In practice, the method of quantities can always be used if $N \geq 50$. For small N, in particular for small σ, it is possible to obtain a negative estimate for σ from formulas (143) as a result of the random fluctuations in the values of τ_i. In this case, the method does not lead to our goal.

Example 6. Objects are tested under a stress $\sigma_{max} = 21$ kg/mm^2 (see Tab. 4.) The longevity of the objects has a distribution with threshold of sensitivity. We wish to estimate the values of c, σ, and t_0.

In this case, the role of τ_i is played by the values of the numbers of cycles N_i. We set $q = 0.15$. We find $\Psi(0.15) = -1.036$ from Tab. A2. By using the rule described, we obtain $\tau_q = 7.95 \cdot 10^5$, $\tau_{1/2} = 17.35 \cdot 10^5$, and $\tau_{1-q} = 112.9 \cdot 10^5$. We calculate $x = 0.098$. From formulas (143), we find $\sigma = 0.97$, $c = 6.02$, and $t_0 = 6.9 \cdot 10^5$. Since the number of experimental data is small, the estimates of

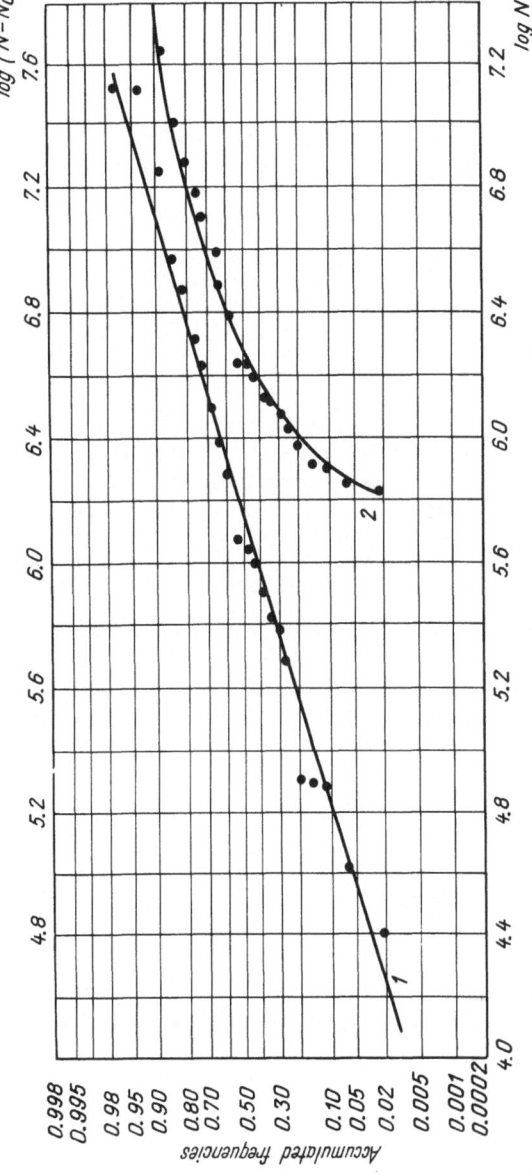

Fig. 42. Smoothing of data on the longevity of objects under low loads on normal-distribution probability paper. The lower scale refers to curve 2, the upper to curve 1. $N_0 = 6 \cdot 10^5$.

the parameters of the distribution are rather crude. This resulted in the value of the threshold $t_0 = 6.9 \cdot 10^5$ turning out to be greater than the smallest of the quantities N_i.

For an approximate estimate of c, σ, and t_0, we can also use normal-distribution paper. Fig. 42 shows the results of smoothing in a logarithmic scale along the abscissa. The initial values of log N_i form a typical bend in the region of small values of N_i (curve 2).

A crude estimate of the value of the threshold of sensitivity is obtained in the following way: we choose a number N_0 less than N_{min} and plot the quantities log $(N_i - N_0)$ along the abscissa. We try to choose the number N_0 in such a way that the accumulated frequencies $\nu(N_i)$ will lie on a straight line. Line 1 of Fig. 42 is obtained by taking the threshold $t_0 = N_0 = 6 \cdot 10^5$ (the upper scale). We see that the points lie reasonably close to a straight line. We can estimate the parameters c and σ by the usual method. Thus, we find from Figure 42 that $c = 6.18$, $\sigma = 0.76$. Keeping in mind the small number of data, we must check to be sure that the results of the calculation do not diverge too much.

It would not be correct to try to shift from a normal to a logarithmic normal distribution every time the histogram of the distribution has an asymmetric form. A logarithmic normal distribution describes the behavior of the lifetime of objects possessing the property of "strenthening" as the time of use increases. This "strengthening" shows up in the constant decrease in the rate of wear. Therefore, before using a logarithmic normal distribution to describe experimental data, it is necessary to use the physical meaning of the process of wear and, if possible, analyze the behavior of the sample functions of the wear to establish whether the objects in question have the property of "strenthening" or not.

Linear sample functions of wear

Up to now, we have discussed only situations in which the initial quality of all the objects is the same and the variations in the lifetime are connected exclusively with the randomness of the process of wear. At the beginning of this section, we pointed out that we often deal with objects for which such a situation does not exist. We again cite the example of the wear (aging) of transistors, the initial quality of which varies considerably and the conditions of use are almost identical and constant. In a number of cases, in bench tests carried out in factories with the purpose of estimating the initial quality of the units, the conditions are intentionally made as nearly constant as possible for all objects tested.

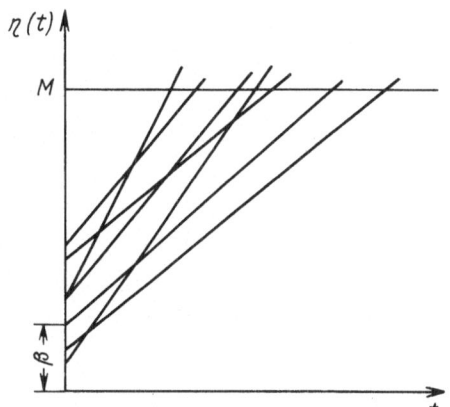

Fig. 43. Linear sample functions of the wear $\eta(t) = \alpha t + \beta$.

In what follows, we confine ourselves to a study of linear wear processes the sample functions of which are straight lines. The general case of linear wear can be analytically represented in the form

$$\eta(t) = \alpha t + \beta, \tag{146}$$

where α and β are independent random variables.

The sample functions are shown in Fig. 43. The quantity β is the initial value of the wear $\beta = \eta(0)$, and the quantity α is the rate or wear. Random variations in α reflect the difference in the initial properties of the object, which cause differing rates of wear. The distribution of α varies according to the form of the object, the conditions under which wear takes place, and so forth. The most natural assumption is that α is the sum of a large number of terms each of which is one component of the rate of wear resulting from the influence of one physical process. For example, we may assume in the case of friction between a shaft and bearing that the rate of wear is the sum of the rates due to abrasion, fatigue, oxidation, and similar processes. If all the terms are of approximately the same order, we may assume that α will follow a normal distribution satisfactorily.

We have already called attention to the fact that a normal distribution gives the variations of a random variable between $-\infty$ and $+\infty$. This means that we allow formally for the possibility that α will be negative. In the processing of experimental data, we can ignore this fact since the probability of negative values of α, which is $\underline{P}\{\alpha \le 0\}$, is usually found to be negligibly small.

If we assume that β also has a normal distribution, then $\eta(t)$ has a normal distribution with parameters

$$\underline{M}\{\eta(t)\} = \underline{M}\{\alpha\}t + \underline{M}\{\beta\}, \tag{147}$$

$$\underline{D}\{\eta(t)\} = t^2 \underline{D}\{\alpha\} + \underline{D}\{\beta\}. \tag{148}$$

If M is the maximum admissible level of wear, then it follows from the obvious equation

$$\underline{P}\{\tau > T\} = \underline{P}\{\alpha T + \beta \le M\} \tag{149}$$

and the normality of $\eta(t)$ that

$$\underline{P}\{\tau \le T\} = 1 - \underline{P}\{\tau > T\} = \Phi\left[\frac{T - \dfrac{M - \underline{M}\{\beta\}}{\underline{M}\{\alpha\}}}{\sqrt{\dfrac{\underline{D}\{\alpha\}T^2 + \underline{D}\{\beta\}}{\underline{M}^2\{\alpha\}}}}\right]. \tag{150}$$

This distribution of the time τ is called Bernstein's distribution [2]. It differs from a normal distribution in that $\underline{D}\{\tau\}$ depends on T. The distribution involves three parameters:

$$a = \frac{\underline{D}\{\alpha\}}{\underline{M}^2\{\alpha\}}, \quad b = \frac{\underline{D}\{\beta\}}{\underline{M}^2\{\alpha\}}, \quad c = \frac{M - \underline{M}\{\beta\}}{\underline{M}\{\alpha\}}. \tag{151}$$

The distribution (150) takes the form

$$F(T) = \Phi\left[\frac{T - c}{\sqrt{aT^2 + b}}\right]. \tag{152}$$

We shall write out a procedure for estimating the parameters a, b, and c that is based on the application of the method of discrimination partitions.

An estimation of the parameters of the distribution (150) from data on the lifetime.

1. We choose numbers T_1, T_2, and T_3 such that $T_1 < T_2 < T_3$ and we calculate the number of values of τ_i that lie in the intervals $(0, T_j)$ for $j = 1, 2, 3$. Suppose that these numbers are equal respectively to $m(T_j)$ for $j = 1, 2, 3$. We calculate the ratios $\nu(T_j) = m(T_j)/N$, where N is the total number of data.

2. From Tab. A2, we find the values $\Psi_j = \Psi(\nu(T_j))$.

3. We use the method of successive approximations or a graphical method to solve the equation

$$\frac{T_1 - T_2}{T_2 - T_3} = \frac{\Psi_1\sqrt{\frac{a}{b}T_1^2 + 1} - \Psi_2\sqrt{\frac{a}{b}T_2^2 + 1}}{\Psi_2\sqrt{\frac{a}{b}T_2^2 + 1} - \Psi_3\sqrt{\frac{a}{b}T_3^2 + 1}} \tag{153}$$

for a/b.

4. From the known ratio $a/b = d$, we find

$$c = \frac{\Psi_2 T_1\sqrt{dT_2^2 + 1} - \Psi_1 T_2\sqrt{dT_1^2 + 1}}{\Psi_2\sqrt{dT_2^2 + 1} - \Psi_1\sqrt{dT_1^2 + 1}}. \tag{154}$$

5. The parameters b and a are determined from the formulas

$$b = \left[\frac{T_1 - c}{\Psi_1\sqrt{dT_1^2 + 1}}\right]^2, \tag{155}$$

$$a = bd.$$

The procedure for estimating the parameters is considerably simplifed if $\beta - \beta_0$ - const. In this case,

$$\underline{D}\{\beta\} = 0,$$

$$F(T) = \Phi\left(\frac{T - c}{\sqrt{a}\ T}\right). \tag{156}$$

1. The data $\tau_1,\ldots,\tau_N$ are partitioned by the chosen numbers T_1 and T_2 into three groups and the values of $\nu(T_1)$ and $\nu(T_2)$ are calculated in a manner analogous to the preceding case. From Table A2, we find the values of $\Psi_i = \Psi(\nu(T_i))$, for $i = 1,2$.

2. To determine the parameters c and a, we have the formulas

$$c = \frac{T_1 T_2 (\Psi_2 - \Psi_1)}{T_2 \Psi_2 - T_1 \Psi_1}, \tag{157}$$

$$\sqrt{a} = \frac{T_2 - T_1}{T_2 \Psi_2 - T_1 \Psi_1}. \tag{158}$$

The case $\beta = \text{const}$ corresponds to a sample function of a fanlike random process, which has the form shown in Fig. 44 (with $\beta = 0$).

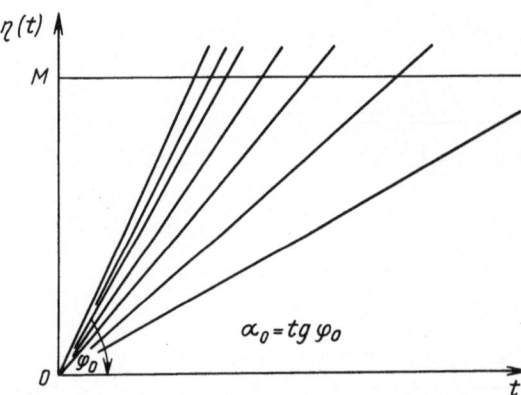

Fig. 44. Linear sample functions of wear $\eta(t) = \alpha t$.

The simplest case is that of a wear process of the type

$$\eta(t) = At + \beta, \tag{159}$$

where A is a constant and β is normally distributed with parameters $\underline{M}\{\beta\}$ and $\underline{D}\{\beta\}$.

The sample function of such a process is shown in Fig. 45. In this case, the parameters of Bernstein's distribution are obviously equal to

$$c = \frac{M - M\{\beta\}}{A}, \quad a = 0, \quad b = \frac{D\{\beta\}}{A^2}. \tag{160}$$

Therefore, the procedure described above for the estimation of three parameters is simplified. Using the partition of the data

$\tau_1, \ldots, \tau_N$ with the quantities T_1 and T_2, we find

$$c = \frac{\Psi_2 T_1 - \Psi_1 T_2}{\Psi_2 - \Psi_1}, \qquad (161)$$

$$\sqrt{b} = \frac{T_1 - c}{\Psi_1}. \qquad (162)$$

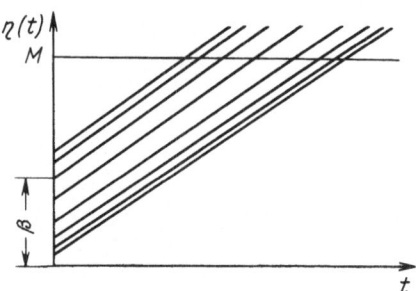

Fig. 45. Linear sample functions of a wear process $\eta(t) = \beta + At$.

Estimation of the parameters of a wear process from sample functions. Suppose that we have at our disposal a certain number of sample functions of wear (it is desirable to have at least 20 or 30). The typical form of such a graph is shown in Fig. 46. We need to estimate the parameters of the distributions of the quantities α and β and to find the probability of failure-free operation dur-

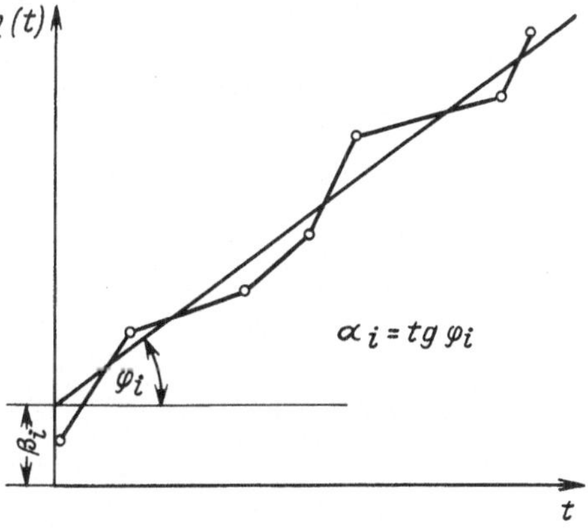

Fig. 46. The estimation of the parameters of a wear process from a sample function.

$$\alpha_i = tg\ \varphi_i$$

ing the course of time T when we have a given limiting level of wear M. The procedure for processing the data reduces to the following:

1. Each of the N sample functions is replaced with a straight line (see Fig. 46). The straight lines are drawn either by the method

of least squares or, if the number of points exceeds 10, they are drawn by eye in such a way that approximately the same number of points will lie above and below the line.

2. For each sample function, we find the value α_i of the tangent of its angle of inclination and the ordinate of its intersection with the β_i-axis.

3. We calculate

$$\underline{M}\{\alpha\} = \frac{\sum\limits_{i=1}^{N} \alpha_i}{N}, \quad \underline{D}\{\alpha\} = \frac{1}{N-1} \sum_{i=1}^{N} (\alpha_i - \underline{M}\{\alpha\})^2, \tag{163}$$

$$\underline{M}\{\beta\} = \frac{\sum\limits_{i=1}^{N} \beta_i}{N}, \quad \underline{D}\{\beta\} = \frac{1}{N-1} \sum_{i=1}^{N} (\beta_i - \underline{M}\{\beta\})^2. \tag{164}$$

4. We find the parameters a,b,and c from formulas (151).

5. We determine the probability of failure-free operation during the time T from the formula

$$\underline{P}\{\tau > T\} = 1 - \Phi\left[\frac{T-c}{\sqrt{aT^2 + b}}\right]. \tag{165}$$

It should be noted that in the case of a small set, the smoothing of experimental data $\tau_1, \tau_2, \ldots, \tau_N$ regarding the lifetime can be done from the distributions of various types with satisfactory agreement. At the same time, the estimates of the probability of failure-free operation during the period of time T that are calculated from different theoretical distributions can give extremely wide deviations. This indicates the necessity of making an analysis of the sample functions and the physical picture of the wear process, especially when we have data on a small number of objects. In conclusion, we note that an estimate of the reliability from the wear sample functions is preferable to making estimates from the lifetime for the following reasons: In the first place, when we have sample functions of the wear process, we can check the agreement between the formal scheme (146) and the actual situation. In the second place, we can find from the sample functions

all the parameters of the wear process (of which there are four),
whereas knowing the parameters a,b, and c of the distribution
(152) does not enable us to do this. Finally, estimation of the
parameters from sample functions is statistically more precise
when other things are equal.

Analysis of sample functions of wear

If we estimate with our eye the behavior of the sample functions
of wear, we can fall into error. At the beginning of this section,
we noted that both the initial quality and the random variations
in the rate of wear have an effect on the behavior of the wear
sample functions. It is very important to try to tell which is the
prevalent factor, and this must be done by objective methods.

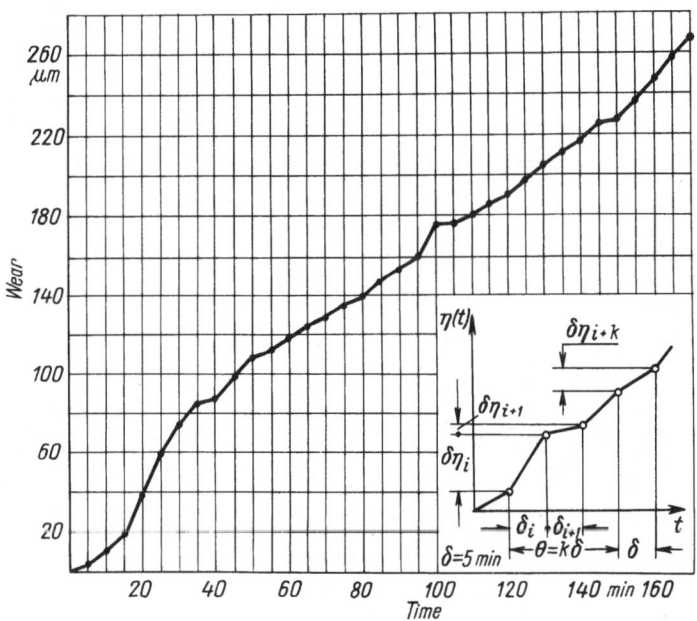

Fig. 47. Sample functions of the wear of retinax.

Consider the sample functions of the wear of a single object
(Fig. 47). The sinuous nature of the sample functions gives us a
basis for assuming that the increments in the wear are independent.

To verify that this is the case, we can indicate a method that is much more reliable than visual analysis.

First of all, we note that complete independence of the increments in the wear is observed rather infrequently. Usually·the increments are dependent, but when the distance θ between the increments is great, the dependence shows up only very weakly, and it decreases with increase in θ. Increments possessing this property are said to be asymptotically independent. We have already used the term "asymptotic independence" in Section 1, when we were discussing the change in the load. Processes with asymptotically independent increments are called processes with strong mixing. This name reflects the physical fact of mixing — that is, the interweaving of the sample functions, of which we spoke above. The assumption of strong mixing of a wear process must be supported in the physics of the phenomena that are taking place. An analysis of the behavior of the sample functions cannot provide direct proofs of this fact. However, it can show whether there are any contradictions between the true and assumed behavior of the sample functions of the wear or not. If the data of the analysis of the sample function contradict the assumption of strong mixing, this assumption must be abandoned.

In what follows, we shall confine ourselves to the case in which the rate of wear is constant on the average. We have already stated that, when the mean rate of wear is nonconstant, we can shift to a constant mean rate by making a change of scale of the t-axis (see p. 73).

One of the most important characteristics of any random process including a wear process is the correlation function. We do not stop here to look at its properties or its theoretical significance but show the correlation function of the increases in wear is calculated and what conclusions can be drawn by studying its behavior [26].

The sample function of the wear shown in Fig. 47 was constructed from the results of measuring the quantity η(t) over equal intervals of time, each of length δ. In each such interval, the value of the wear assumes a certain random increment $\delta\eta_i$. If we take two adjacent increments $\delta\eta_i$ and $\delta\eta_{i+1}$, we can expect them to prove dependent. If we take two increments $\delta\eta_i$ and $\delta\eta_{i+k}$, the connection between them may be very weak for large values of k. It is desirable to have a measure for quantitative estimate of this connection. In mathematical statistics, one generally uses the correlation coefficient as a measure of the connectedness of two random variables. Let x and y denote two random variables, the observed values of which are given in the form of Tab. 5.

Table 5. The values of the random variables x and y.

x	x_1	x_2	x_3	...	x_{n-1}	x_n
y	y_1	y_2	y_3	...	y_{n-1}	y_n

Under each value of the random variable x appears the corresponding value of the random variable y.

The correlation coefficient between the random variables x and y is calculated in accordance with the formula

$$r_{xy} = \left(\frac{1}{n}\sum_{i=1}^{n} x_i y_i - \overline{xy}\right)\Big/ s_x \cdot s_y, \qquad (166)$$

where

$$\overline{x} = \frac{1}{n}\sum_{i=1}^{n} x_i; \qquad \overline{y} = \frac{1}{n}\sum_{i=1}^{n} y_i; \qquad (167)$$

$$\varepsilon_x = \sqrt{\left[\sum_{i=1}^{n}(x_i - \overline{x})^2\right]\Big/ n}; \qquad s_y = \sqrt{\left[\sum_{i=1}^{n}(y_i - \overline{y})^2\right]\Big/ n}. \qquad (168)$$

The quantities $\overline{x}$ and $\overline{y}$ are the empirical means of the random variables x and y; s_x and s_y are their standard deviations.

Example 7. The quantities $\delta\eta_i$ corresponding to the sample function shown in Fig. 47 are given in Tab. 6. Find the correlation coefficient to the variables $\delta\eta_i$ and $\delta\eta_{i+1}$, which are shown one above the other in Tab. 6 along the pattern of Tab. 5. We note that, in Tab. 6, the same increment appears twice — in the upper and in the lower row. The only exceptions are the first and last increments.

In accordance with formulas (167) and (168), we obtain

$$\overline{x} = 8.12, \quad \overline{y} = 8.27, \quad s_x = 4.71, \quad s_y = 4.62, \quad \frac{1}{n} \sum_{i=1}^{n} x_i y_i = 76.45.$$

$$r_{xy} = 0.43.$$

Table 6.

Increments in the wear of the sample function shown in Figure 47.

j	1	2	3	4	5	6	7	8	9	10	11	12	13	14	15	16	17
$\delta\eta_j$	3	7	8	20	21	15	11	2	12	9	4	6	5	5	7	3	10
$\delta\eta_{j+1}$	7	8	20	21	15	11	2	12	9	4	6	5	5	7	3	10	6

continuation of Table 6)

j	18	19	20	21	22	23	24	25	26	27	28	29	30	31	32	33	34
$\delta\eta_j$	6	6	16	0	4	6	4	8	7	7	6	7	2	9	11	11	8
$\delta\eta_{j+1}$	6	16	0	4	6	4	8	7	7	6	7	2	9	11	11	8	

The correlation coefficient is never greater in absolute value than unity. Also, $|r_{xy}| = 1$ if there ia a linear relation between the random variables x and y. On the other hand, if x and y are independent, then $r_{xy} = 0$. Therefore, the value of the coefficient r_{xy} serves to some extent as a measure of the functional connection of the random variables. The rather large value of r_{xy} obtained in the example above indicates that neighboring increments in the wear are connected. In analogy with Tab. 6, one can construct tables of the interrelationship acquired over a distance $\theta = k\delta$ for $k = 0,1,2$ with each other. We denote by $r(\theta)$ the value of the correlation coefficient of the increments η_i, η_{i+k} that lie at a distance $\theta = k\delta$ from each other. We note that, for $\theta = 0$, the coefficient $r(0)$ characterizes the relationship between the increments η_i. Therefore, $r(0)$ is always 1. For $\theta = \delta$, the coefficient $r(\delta)$ characterizes the relationship between neighboring increments, etc. (see Fig. 47).

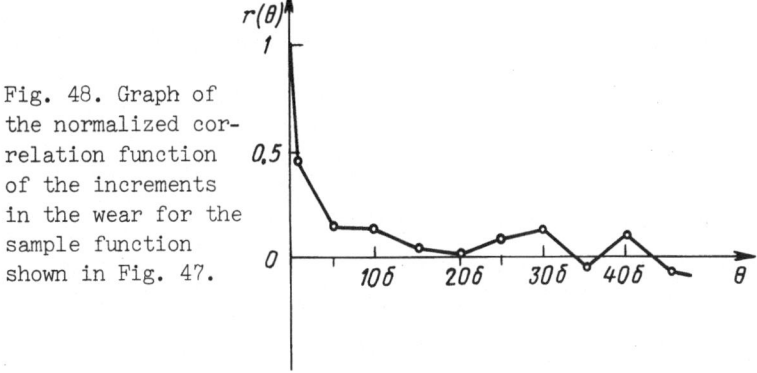

Fig. 48. Graph of the normalized correlation function of the increments in the wear for the sample function shown in Fig. 47.

Fig. 48 shows the graph of the function $f(\Theta)$ constructed from the sample function of Fig. 47 and its continuation on the interval from 180 to 500 min. It should be noted that, for a relatively reliable estimate of the values of $r(\Theta)$, we need to have at least 70 to 100 values of $\delta\eta_i$. The function $r(\Theta)$ is called the <u>normalized correlation function</u>. We point out some important features

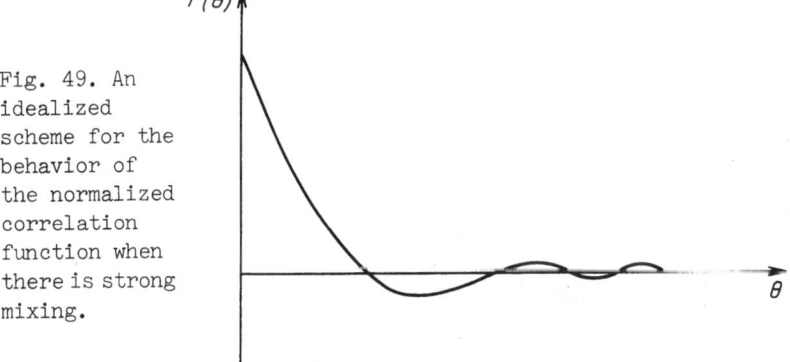

Fig. 49. An idealized scheme for the behavior of the normalized correlation function when there is strong mixing.

of this graph: for small values of Θ, the correlation function is rather large. It then decreases sharply and oscillates about zero over a fairly wide range. The correlation function of the increments must behave in just this way when the process of wear possesses the property of strong mixing, that is, the property of asymptotic independence of the increments. Fig. 49 shows an idealized scheme of the behavior of the normalized correlation function of the increments when this is the case. The characteristic feature is the fact that the correlation function tends to 0 as the distance Θ between the increments increases.

An important question deals with the general continuability of observation of the course of change in the wear. Usually, the lifetime determined by natural wear is extremely great. Observation of the course of change in the wear during this entire time is technologically difficult and expensive. Such a protracted observation of the wear is not necessary even in the plan of study of the correlation function. If we wish to estimate the probability of failure-free operation during a period of time T, it will be sufficient to study the behavior of the wear during an interval of time of length approximately $(0.1 - 0.2)T$ with the stipulation, however, that the basic time of observation occupy the zone of normal wear. For a comparatively brief period of time of observation of the course of wear, there is the danger that the nature of the behavior of the wear may change in the period following the period of observation. Therefore, the extrapolation into the fu-

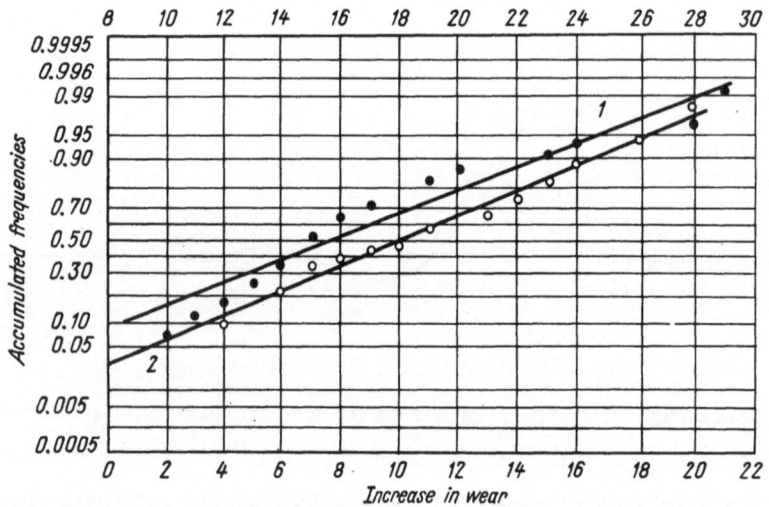

Fig. 50. Smoothing of the increments of wear. Line 2 corresponds to the upper scale.

ture may be erroneous. Here, a preliminary experiment on the study of analogous machines and devices is important. By simply resting on the experimental material at our disposal, we can tell with a fair degree of safety whether the extrapolation of the data with respect to the change in wear is justified or not.

For example, it is possible to extrapolate the data with respect to the change in steepness of the characteristic of a vacuum tube. The assurance of the possibility of such extrapolation is based on the behavior already studied of the curvature of a large number of vacuum tubes of various types.

In addition to the correlation function, it is important to study the distribution of the quantities $\delta\eta_i$, that is, the increments in the wear. If a process of wear possesses the property of strong mixing, then the distribtuion of the variables $\delta\eta_i$ must be approximately normal for a rather long period of time δ. [1]

Fig. 50 shows the results of smoothing the values of the increments in the wear over an interval of $\delta = 5$ min (the straight line 1) and over an interval of $\delta = 15$ min (line 2). The data were chosen from the complete sample function observed on an interval of length 500 min. It is obvious from the drawing that the distribution of the increments approximates a normal distribution with increase in the interval δ.

An analysis of the single sample function of the wear enables us to judge whether hypothesis of strong mixing of the wear process is valid or not by using experimental data. However, from this analysis we cannot answer the question as to whether the initial quality of the random fluctuations in the state of the objects is prevalent in the behavior of the sample functions. To answer this question, we must have several sample functions at our disposal. Before explaining the sequence used in the analysis of the sample functions, let us indicate the peculiarities of the behavior of

---

[1] The properties of processes with strong mixing have been intensively studied in recent years. The basic properties of these processes are clarified in the works by A. N. Kolmogorov, I. A. Ibragimov, and Yu. A. Rozanov.

the variance $\underline{D}\{\eta(t)\}$ when the sample functions of the wear are linear and there is strong mixing. In the first place, the wear process is given by the formula

$$\eta(t) = \alpha t + \beta. \tag{169}$$

When we calculate the variance, we obtain

$$\underline{D}\{\eta(t)\} = \underline{D}\{\alpha\}t^2 + \underline{D}\{\beta\}. \tag{170}$$

In the theory of processes with strong mixing, it is shown that $\underline{D}\{\eta(t)\}$ increases linearly, that is, that

$$\underline{D}\{\eta(t)\} = At + B, \tag{171}$$

where A and B are constants.

Thus, if the behavior of the sample functions of the wear is determined entirely by the initial quality, then the variance $\underline{D}\{\eta(t)\}$ will increase like t^2 and, when there is strong mixing and the initial quality is homogeneous, it will increase like t. In view of what has been said, it is obvious that in analyzing the sample functions of the wear, we need to study the behavior of the variance $\underline{D}\{\eta(t)\}$.

To be certain of the nature of the wear, we need to have at our disposal at least 25 or 30 sample functions and 30 to 50 points of measurement of $\eta(t)$ on each of them.

We recommend the following procedure for analyzing the sample functions of the wear. (It is assumed that the observations of the wear have been made for all sample functions at the same instants of time t_j, for $j = 1, 2, \ldots, m$.)

1. We construct a mean line for the entire family of sample functions. To do this, we calculate the mean value of the wear $\overline{\eta}(t_j)$ at each instant of time t_j from the formula

$$\overline{\eta}(t_j) = \frac{1}{N} \sum_{i=1}^{N} \eta^{(i)}(t_j), \tag{172}$$

where N is the total number of sample functions and $\eta^{(i)}(t_j)$ is the wear according to the ith sample function at the instant t_j. If the mean rate is constant, the dependence of $\overline{\eta}(t_j)$ on t_j must be linear.

100

2. For every instant t_j, the variance $s^2_\eta(t_j)$ is calculated according to the formula

$$s^2_\eta(t_j) = \frac{1}{N-1} \sum_{i=1}^{N} (\eta^{(i)}(t_j) - \overline{\eta}(t_j))^2. \tag{173}$$

The values of the variances are plotted on a graph. If they lie close to a straight line, this fact can serve as an indication of homogeneity of the quality of the objects. For our estimate of the behavior of the variance to be objective, it is useful to smooth the quantities $s^2_\eta(t_j)$ in accordance with a quadratic dependence of the form

$$s^2_\eta(t) = a_0 + a_1 t + a_2 t^2. \tag{174}$$

The coefficient a_2 in this relation indicates the influence of the difference between the objects in the initial state on the variance, and the coefficient a_1 indicates the influence of the random fluctuations in the states of the objects during the course of their use. Within the limits of a given time t for which we are estimating the probability of failure-free operation, the influences of each of these two factors can be evaluated by comparing the quantities $a_1 t$ and $a_2 t^2$. If the quantity $a_2 t^2$ is small in comparison with $a_1 t$, then the last term of formula (174) can be neglected. Hence, we may assume that the variance increases linearly and that the influence of the initial quality is weak. If the quantity $a_1 t$ is small in comparison with $a_2 t^2$, we can neglect the randomness of the fluctuations in the states of the object, assuming that the sample functions are ruled.

3. For each individual sample function of the wear, we write out the row of increments in the wear of the type of the first row of Tab. 6. These rows "adjoin" into a single general series as indicated in Tab. 7 (the first row for $\delta\eta_j$). The second, third, and other rows of Tab. 7 contain the increments $\delta\eta_{j+1}$, $\delta\eta_{j+2}$, etc., displaced in each case one step to the left, taken from the first row. In calculating $r(\delta)$, we put in juxtaposition the values of the increments of the first and second rows, in calculating $r(2\delta)$,

Table 7   Combined table of increments in the wear for the group of sample functions

	1st sample function ($i=1$)					2nd sample function ($i=2$)					$\cdots$	Nth sample function ($i=N$)				
$\delta\eta_j$	$\delta\eta_1^{(1)}$	$\delta\eta_2^{(1)}$	$\delta\eta_3^{(1)}$	$\cdots$	$\delta\eta_m^{(1)}$	$\delta\eta_1^{(2)}$	$\delta\eta_2^{(2)}$	$\delta\eta_3^{(2)}$	$\cdots$	$\delta\eta_m^{(2)}$	$\cdots$	$\delta\eta_1^{(N)}$	$\delta\eta_2^{(N)}$	$\delta\eta_3^{(N)}$	$\cdots$	$\delta\eta_m^{(N)}$
$\delta\eta_{j+1}$	$\delta\eta_2^{(1)}$	$\delta\eta_3^{(1)}$	$\cdots$	$\delta\eta_m^{(1)}$		$\delta\eta_2^{(2)}$	$\delta\eta_3^{(2)}$	$\cdots$	$\delta\eta_m^{(2)}$		$\cdots$	$\delta\eta_2^{(N)}$	$\delta\eta_3^{(N)}$	$\cdots$	$\delta\eta_m^{(N)}$	
$\delta\eta_{j+2}$	$\delta\eta_3^{(1)}$	$\cdots$	$\delta\eta_m^{(1)}$			$\delta\eta_3^{(2)}$	$\cdots$	$\delta\eta_m^{(2)}$			$\cdots$	$\delta\eta_3^{(N)}$	$\cdots$	$\delta\eta_m^{(N)}$		
$\vdots$	$\vdots$	$\vdots$	$\vdots$	$\vdots$	$\vdots$	$\vdots$	$\vdots$	$\vdots$	$\vdots$	$\vdots$	$\vdots$	$\vdots$	$\vdots$	$\vdots$	$\vdots$	$\vdots$

$r(\delta)$ $r(2\delta)$

$\delta\eta_j^{(i)}$ is the increment in the wear calculated from the ith sample function during the jth interval of the time of observation.

we put in juxtaposition the values of the increments of the first
and third rows, and so on. We observe that, in calculating the
correlation function $r(\Theta)$, the quantities $\overline{x}, \overline{y}, s_x$, and s_y, calculated from formulas (168) and (167) must be obtained without sep-
arating the sample functions, that is, for the entire row of the
table. When we have a large number of data, the calculations become rather laborious. Therefore, it is expedient to use present-
day computing machines to carry out the calculations.

If there is strong mixing, the graph of $r(\Theta)$ must recall Fig. 49.
4. If the increments have a normal distribution, this fact can
serve as an extra argument that there is strong mixing. To see
whether this distribution is normal or not, we smooth the set of
increments (first row of Tab. 7) on normal-distribution paper. If
the data do not lie approximately on a straight line, we must make
a smoothing of the increments in the wear calculated over a wide
interval, for example, the interval $2\delta, 3\delta$, or the like.

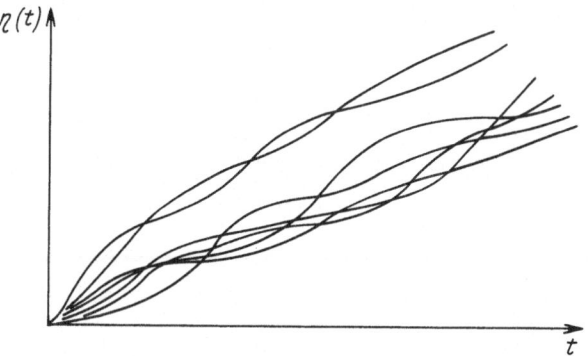

Fig. 51. The form of the sample functions of the wear when certain objects are of poor quality.

The most complicated situation is one in which the terms $a_1 t$ and
$a_2 t^2$ in formula (174) are of the same order of magnitude. The cir-
cumstances that lead to this situation can be quite varied. Thus,
it sometimes happens that the majority of the sample functions indicate homogeneity of the initial quality though certain objects
are of noticeably poorer quality. The sample functions of the wear
of "poor" objects are separated from the remaining ones (see Fig.

51). We can also have the picture shown in Fig. 52, where it is difficult to separate certain sample functions from the others although the influence of the difference in the initial quality of the objects is perceptible. Therefore, whenever the quantities $a_1 t$ and $a_2 t^2$ are about the same size, we need to make a careful survey of the sample functions before proceeding with subsequent calculations.

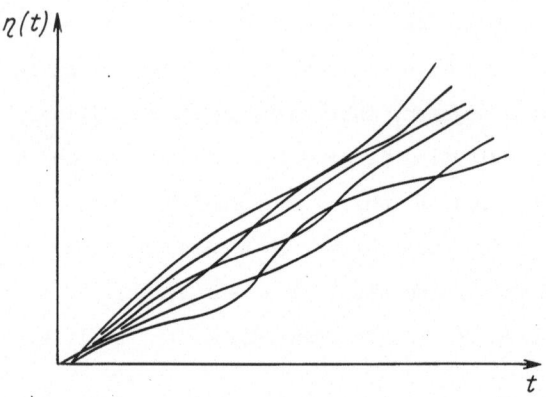

Fig. 52. The form of the sample functions of the wear when the initial quality and the presence of mixing have a strong influence.

If certain of the sample functions can be separated from the others, we need to repeat the numerical analysis of the sample functions according to the method described above preliminarily excluding these from consideration. When we have satisfied our-selves that the remaining sample functions satisfy the hypothesis of homogeneity of the initial quality, we can make a separate cal-culation of the reliability for the "satisfactory" objects. The presence of individual low-quality objects must either be allowed for or numerically estimated. To get a numerical estimate of the influence of the "poor" objects, we need to calculate the portion of them out of all N of the sample functions studied. Suppose that this portion is equal to q, so that qN sample functions lie above the remaining homogeneous $(1 - q)N$ ones (see Fig. 51). For the "poor" sample functions we can estimate individually the proba-bility of failure-free operation. This estimate will be extremely crude since the number of such sample functions will be small. If the "poor" sample functions are interwoven, we need to use the method described on page 51; if they are separated from each other, we should use the method described on page 91. Let $F_1(T)$ denote the distribution function of the time τ for the "satisfactory" sample functions, and let $F_2(T)$ denote the distribution function of the time τ for the "poor" ones. Then, for the entire set of objects, we obtain the distribution function

$$F(T) = (1 - q)F_1(T) + qF_2(T),$$

which yields a superposition of the two distributions $F_1(T)$ and $F_2(T)$.

Thus, we have arrived at the superposition of two normal distributions of the lifetime τ. On page 25, we listed reasons leading to a superposition of exponential distributions. One can easily see that a superposition of distributions will occur whenever several groups of objects of different quality are mixed.

If the sample functions that we are studying are such that, first, neither of the quantities $a_1 t$ and $a_2 t^2$ can be neglected, and, second, we cannot single out the group of "poor" sample functions, then a crude estimate of the reliability can be obtained from the formula

$$\underline{P}\{\tau > T\} = \Phi\left[\frac{M - \overline{\xi}T}{\sqrt{a_0 + a_1 T + a_2 T^2}}\right], \tag{175}$$

where the constants a_0, a_1, and a_2 are the coefficients in equation (174), $\overline{\xi}$ is the mean rate of wear, and M is the maximum admissible level of wear.

In conclusion, we emphasize once more that the analysis of the sample functions must be regarded as an auxiliary means of verifying the hypothesis of the physical picture of a failure. An examination of the cause of failure, a detailed study of the reasons for a drop in reliability, and a careful evaluation of the influence of factors associated with production and use must be made when one is calculating the reliability. Only then can this calculation serve to develop measures for increasing the quality and reliability of the units.

Nonetheless, it would be incorrect to assert that we have to have extremely precise information regarding the form of the distribution law of the lifetime τ every time we make a calculation of one of the numerous indices of reliability or, in general, whenever we make a calculation of devices that is to some degree connected with the reliability of its elements. In fact, in a number of calculations, our information regarding the distribution of the time τ can be extremely approximate and yet this fact does not actually show up in the accuracy of the calculation. Therefore, we need to single out the bare requirements of this or that calculation of the reliability and the general plan of investigating the reliability. On page 155, we give a reference table with an indication of the necessary degree of knowledge of the distribution of the time τ depending on the nature of the calculation of the reliability. Situations in which the distribution of the time τ must be known with a high degree of accuracy and certainty are comparatively rare. But a penetrating study of the essence of failure discloses reserves of increase in reliability and, for this reason, it is sometimes the decisive factor in the general plan of investigation.

Chapter IV

Processes of Damage with Relaxation

In this chapter, we shall study a model of the occurrence of a
failure that, in a certain sense, is more general than the model
of instantaneous injury or the model of accumulation of injuries.
Let us first look at some examples.

We know that in electro-automatic systems, breakdowns frequently
occur in the insulation of solenoidal coils, relays, etc. When we
were studying production lines for relay springs, we noted many
cases of burning out of the coils of solenoidal presses. With repeated work cycles, there appeared a gradual ever-increasing
"build-up" of the coil frame and a wearing-down of the ring of the
solenoid. With increase in the clearance, the probability that the
windings of the coil will approach each other and their insulation will be rubbed off increases. And this proves to a certain
degree a reason for failure.

In mechanical systems, we frequently encounter failures of the
type of wedging of moving parts in the controls. An analysis of
specific cases shows that the deciding role is played by the accumulation of wear and of clearances in the conjugate kinemetic
pairs, which leads to an increase in the probability of wedging.
A breakdown in the operative memory of an electronic computing
machine can occur if the impulse of the current fed to the coor-
dinate transformer is insufficient for its magnetic reversal. The
change in the form of the impulses (decrease in height, decrease
in steepness, etc.) occurs gradually and randomly under the influence of a change primarily in the parameters of the vacuum
tubes as a result of aging, falling of the slope of their characteristics, etc. As the form of the impulse is distorted, the work

becomes unstable and the probability of a breakdown, that is, an incorrect writing of information in the memory cell, increases. Without going into the particular nature of these failures, we can note the following general peculiarities regarding their occurrence. The element (or system) has some parameter $\eta(t)$ that changes in the course of operation as a random function of time (usually either nondecreasing or nonincreasing). The change in the parameter $\eta(t)$ does not itself cause failure. Failure occurs, as it were, in another place but it is connected with the change in the parameter in such a way that increase (decrease) in the parameter causes an increase (decrease) in the probability of occurrence of a failure. The failure occurs in the form of a step change in the state of the system (breakdown, wedging, incorrect writing of information in a cell), but this is preceded by a process of accumulation of injuries. In the Introduction, we refer to such a scheme of occurrence of failure as relaxation. This failure model has the features both of a model of instantaneous injury (step change in the state) and of a model of accumulation of injuries (gradual increase, accumulation of injuries with increase in the probability of failure). In more complicated cases, the failure occurs as the result not of one but of many relaxations.

The central feature of a process of injury with relaxation is the fact that the parameter of the system $\eta(t)$ which affects the probability of failure (it is convenient to call this parameter the prognosticating parameter) changes in an indeterminate way, that is, randomly. This last circumstance admits a formal description of the following form: the probability of failure of the system or element in the interval $(T, T + \Delta_T)$ when the prognosticating parameter is equal to $\eta(T)$ up to the instant T is determined by

$$\gamma(T) = \varphi[\eta(T)]\Delta_T + o(\Delta_T), \qquad (176)$$

where $\eta(t)$ is the prognosticating parameter and φ is some function.

To make this model more specific, let us make the following most natural assumptions. Let us assume that $\eta(t)$ is a process of wear corresponding to a scheme of accumulated injuries (see pp. 38, 55) and that the function φ is linear:

$$\varphi = A + B\eta. \tag{177}$$

To derive the distribution law of the lifetime, we resort to the following interpretation of a model of occurrence of failure.

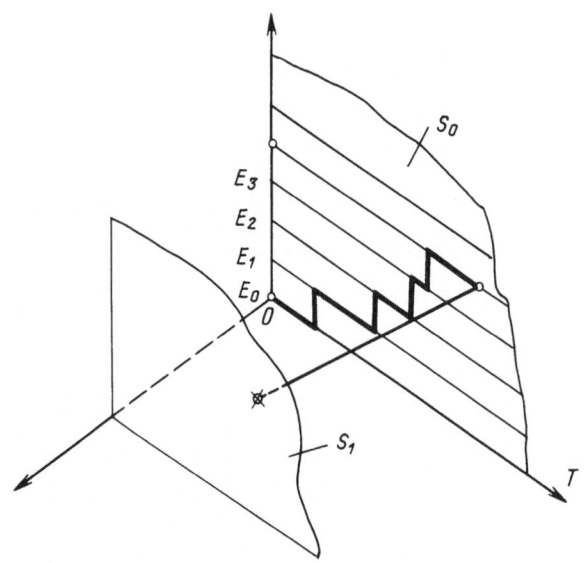

Fig. 53. A model of the occurrence of a failure under random change in the prognosticating parameter.

Let us consider the spatial wandering of a fictitious particle (see Fig. 53). The wandering begins at the instant $T = 0$ from a point 0 in the plane S_0. At random instants of time, the particle in the plane S_0 actually describes the process of accumulation of injuries. The probabitlity of receiving a single injury, that is, the probability of the shift $E_k \to E_{k+1}$ in the interval $(T, T + \Delta_T)$ is equal to $\lambda\Delta_T + o(\Delta_T)$ (cf. the model of occurrence of the gamma distribution, p. 55).

For each level E_k, there is a probability that the particle will jump onto the wall S_1, which is an absorbing screen. This jump of the particle onto S_1 simulates a failure of the system. The dependence of the probability of failure on the value of the parameter is reflected in the fact that the probability of shifting from the level E_k to S_1 during the interval $(T, T + \Delta_T)$ depends on k and is equal to

$$q_k(T) = [\mu_0 + k\mu]\Delta_T + o(\Delta_T). \qquad (178)$$

At every instant of time T, the state of the particle can be characterized by giving the values of E_j and S_i for $j = 0, 1, \ldots$ and
$i = 0, 1$, that is, by giving the height of the level and the number
of the wall that it is on. The probabilities of such a shift during a small interval of time $(T, T + \Delta_T)$ are determined on the basis of the assumptions made as follows (see Tab. 8):
Suppose that $P_{j,i}(T)$ is the probability that the particle will be
in the state $(\underline{E}_j, \underline{S}_i)$ by the instant T. We can set up the following
infinite system of equations for the functions $\{P_{k,0}(T), k = 0, 1, \ldots\}$:

$$
\left.
\begin{aligned}
&P_{0,0}(T + \Delta_T) = P_{0,0}(T)[1 - (\mu_0 + \lambda)\Delta_T] + o(\Delta_T), \\
&\cdots\cdots\cdots\cdots\cdots\cdots\cdots\cdots\cdots\cdots\cdots \\
&P_{k,0}(T + \Delta_T) = P_{k,0}(T)[1 - (\mu_0 + \mu k + \lambda)\Delta T] + \\
&+ P_{k-1,0}(T)\lambda\Delta_T + o(\Delta_T). \\
&k = 1, 2, \ldots
\end{aligned}
\right\} \qquad (179)
$$

In fact, the particle can be in the state $(\underline{E}_k, \underline{S}_0)$ at the instant
$T + \Delta_T$ only in two incompatible ways: first, it may be in the state
$(\underline{E}_k, \underline{S}_0)$ at the instant T and does not go anywhere during the in-

Table 8. Probabilities of shifts

The ini- tial state	The final state	The probability of a shift during the interval $(T, T + \Delta_T)$
$(\underline{E}_j, \underline{S}_0)$	$(\underline{E}_j, \underline{S}_1)$	$(\mu_0 + \mu j)\,\Delta_T + o(\Delta_T)$
$(\underline{E}_j, \underline{S}_0)$	$(\underline{E}_{j+1}, \underline{S}_0)$	$\lambda\Delta_T + o(\Delta_T)$
$(\underline{E}_j, \underline{S}_0)$	$(\underline{E}_j, \underline{S}_0)$	$1 - (\lambda + \mu_0 + \mu j)\,\Delta_T + o(\Delta_T)$

terval $(T, T + \Delta_T)$; second, it can be in the state $(\underline{E}_{k-1}, \underline{S}_0)$ at the instant T and it shifts to the state $(\underline{E}_k, \underline{S}_0)$ at some time between T and $T + \Delta_T$.

After making some transformations, dividing both sides of the kth equation by Δ_T, and taking the limit as $\Delta_T \to 0$, we obtain the following system of differential equations:

$$
\left.
\begin{aligned}
&P'_{0,0}(T) = - [\mu_0 + \lambda] P_{0,0}(T), \\
&\cdots\cdots\cdots\cdots\cdots\cdots\cdots\cdots\cdots\cdots \\
&P'_{k,0}(T) = - [\lambda + \mu_0 + k\mu] P_{k,0}(T) \\
&\qquad\qquad + \lambda P_{k-1,0}(T), \\
&k = 1, 2, \ldots
\end{aligned}
\right\}
\tag{180}
$$

The initial condition for this system is the condition

$$
P_{i,0}(0) =
\begin{cases}
1, & \text{if } i = 0, \\
0, & \text{if } i \neq 0,
\end{cases}
\tag{181}
$$

which follows from the fact that the particle is in the state $(\underline{E}_0, \underline{S}_0)$ at the initial instant $T = 0$.

We define

$$
R(T) = \sum_{j=0}^{\infty} P_{j,0}(T).
\tag{182}
$$

In the sense of the probabilities $\underline{P}_{j,0}(T)$, the quantity $R(T)$ is the probability that the particle will at the instant T be in some state (the zeroth, first, etc.) on the wall $\underline{S}_0$, that is, the probability that the random time τ of its wandering exceeds T. This means that we are justified in writing

$$
\underline{P}\{\tau \leq T\} = F(T) = 1 - R(T).
\tag{183}
$$

This last relationship defines the distribution function of the time before a jump (failure) in the model that we are considering. The system (180) can be solved recursively and, by carrying out the summation, we can find $R(T)$ and $F(T)$. We omit the calculation and present the final result:

$$
F(T) = 1 - \exp\left[\frac{\lambda}{\mu} - (\lambda + \mu_0)T - \frac{\lambda}{\mu} \exp(-\mu T)\right].
\tag{184}
$$

The corresponding distribution density is equal to

$$f(T) = \{\mu_0 + \lambda[1 - \exp(-\mu T)]\} \times$$

$$\times \exp\left\{\frac{\lambda}{\mu}[1 - \exp(-\mu T)] - (\lambda + \mu_0)T\right\}, \tag{185}$$

and the failure rate is equal to

$$\lambda(T) = \mu_0 + \lambda(1 - e^{-\mu T}). \tag{186}$$

The form of the density curves of this distribution is shown in Fig. 54. The ordinate of the density at zero is equal to the parameter μ_0. For large values of μ_0, the function $f(T)$ decreases monotonically. When μ_0 is small in comparison with λ and μ, the density curve has a "hump" the vertex of which lies to the left of the mathematical expectation.

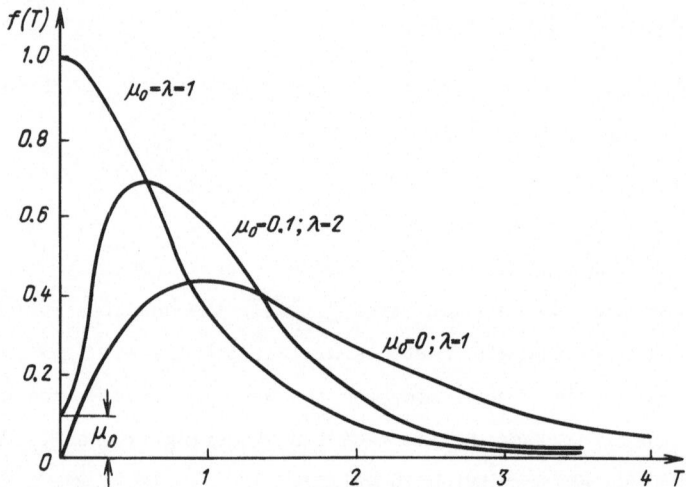

Fig. 54. The density of the distribution (185) for $\mu = 1$ and various values of the parameters λ and μ_0.

Let us analyze the assumptions made in the choice of the distribution (184). The first assumption regards the nature of the change in the prognosticating parameter $\eta(T)$. It amounts to the assumption that the process $\eta(T)$ corresponds to the case of homogeneous initial quality of the objects, a constant mean rate of wear, and the existence of random variations in the rate for each sample function (see p. 38). The second assumption is that the probability of failure depends linearly on the value of the prognosticating parameter. At first glance, this assumption does not seem to have any clear physical meaning.

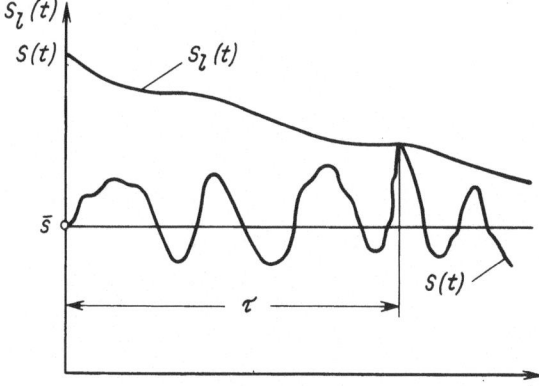

Fig. 55. A random load with decreasing admissible level.

We mention one frequently encountered situation for which a linear relationship between the probability of failure and the level of the prognosticating parameter is typical. Let us return to the case of a failure that occurs when the working load $S(t)$ exceeds the maximum admissible level $S_1(t)$. Suppose that the load $S(t)$ is a normal stationary random process with mean level $\bar{s}$ and that $S_1(t)$ decreases monotonically and randomly under the influence of a randomly proceeding process of aging of the system (see Fig. 55). In this case, the probability of failure depends on the "distance" between $S_1(t)$ and $\bar{s}$, and the quantity $S_1(0) - S_1(t) = \Delta S_T$ takes the role of prognosticating parameter. It can be shown [1] that if $S_1(T) = S_T$ at the instant $t = T$, then the probability of failure in the interval $(T, T + \Delta_T)$ is equal to

$$\gamma(T) = C\, e^{-g(S_T - \bar{s})^2} \Delta_T + o(\Delta_T), \tag{187}$$

where g and C are constants.

Now, let us suppose that $S_1(t)$ changes slowly with the passage of time, that is, that the probability of attaining the maximum ad-

[1] For a normal random process and its intersections with the level S_1, the reader is referred to [14].

missible level during a period of time comparable to the mean time prior to a failure is small. In this case, the factor in front of Δ_T in formula (187) can be represented in the form

$$C\,e^{-g l_0^2}[1 + 2g l_0 \Delta S_T], \qquad (188)$$

where

$$l_0 = S_1(0) - \bar{s}.$$

From this it is obvious that the probability of failure depends linearly on the value of the prognosticating parameter ΔS_T. Let us pause to estimate the parameters of the distribution (184). If we have at our disposal data regarding the lifetime $\tau_1, \tau_2, \ldots, \tau_N$ of N units and if we can observe the sample functions of the prognosticating parameter $\eta(t)$, the following estimating procedure is recommended. From the sample functions of $\eta(T)$, we find the parameter λ just as in the section on the gamma distribution.

From the data $\tau_1, \ldots, \tau_N$, we construct the histogram. The value μ_0 is taken equal to the ordinate of the histogram at the point $T = 0$. We then partition the data by means of a number Θ and we calculate the frequency $\nu(\Theta) = m(\Theta)/N$, where $m(\Theta)$ is the number of cases such that $\tau_i < \Theta$.

The value of μ is found by solving the equation (graphically, for example)

$$\frac{1 - \exp(-\mu\Theta)}{\mu} = \Theta\left(1 + \frac{\mu_0}{\lambda}\right) + \frac{\ln[1 - \nu(\Theta)]}{\lambda}. \qquad (189)$$

On the right side of (189) is a known quantity; the left side decreases monotonically from Θ to 0 as μ increases.

If it is not possible to observe the sample functions $\eta(t)$ but we have data on the lifetime $\tau_1, \tau_2, \ldots, \tau_N$, then to get an estimate for the parameters we can use the method of discrimination partitions.

Let $\Theta_1, \Theta_2, \Theta_3$ denote points of partition such that $\Theta_1 < \Theta_2 < \Theta_3$. We define

$$\frac{\lambda}{\mu} = a,$$

$$\lambda + \mu_0 = b,$$

$$\ln[1 - \nu(\Theta_i)] = \alpha_i.$$

Then, if we take $F(\Theta_i) = \nu(\Theta_i)$, we obtain a system of three equations for finding $a, b,$ and μ:

$$a(1 - e^{-\mu\Theta_i}) - b\Theta_i = \alpha_i \qquad i = 1, 2, 3. \tag{189a}$$

By means of some simple transformations, we obtain an equation in μ:

$$\frac{\varphi(\mu, \Theta_1) - \varphi(\mu, \Theta_2)}{\varphi(\mu, \Theta_1) - \varphi(\mu, \Theta_3)} = X,$$

where

$$\varphi(\mu, \Theta) = \frac{1 - e^{-\mu\Theta}}{\Theta}; \qquad X = \frac{\dfrac{\alpha_1}{\Theta_1} - \dfrac{\alpha_2}{\Theta_2}}{\dfrac{\alpha_1}{\Theta_1} - \dfrac{\alpha_3}{\Theta_3}}.$$

After we have found μ, the quantities a and b can be determined by using two equations in the system (189a).

Chapter V

Superposition of Causes of Failures. Chain Systems

The simultaneous action of several causes of failure

In the preceding chapters, we considered the interaction of "peak"
loads and the process of wear. The action of wear showed up in the
constant lowering of the maximum admissible load level, and the
exceeding of this level by a "peak" in the load led to failure.
Thus, the wear and the instantaneous injury leading to failure
depend on the maximum admissible load. However, we rather frequently observe objects where the wear and the failures due to
"peak" loads are unconnected with each other. Aviation equipment
is an example. The gradual wear of such equipment does not lead
to a drop in their durability under overloading. This is explained
by the fact that the parts of the devices detecting the overloadings, to all intents and purposes, do not wear out. But wear
can cause a violation of its capacity to function by causing the
operating characteristic $\eta(t)$ to get outside admissible limits.
Overloadings, in turn, can cause a failure if their value exceeds
the calculated limits. Consequently, the "peak" loads, which appear
here in the form of overloadings, and wear, which affect the behavior of the operating characteristic $\eta(t)$, act parallel to but independently of each other. Such examples could be multiplied.
Actually, any object that has several operating parts that do not
interact with each other also has several causes of failure that
act parallel to but independently of each other.
Suppose that we have several causes $\underline{\Pi}^{(1)}, \underline{\Pi}^{(2)}, \ldots, \underline{\Pi}^{(k)}$, that act
parallel to but independently of each other. Their independence
means that the action of any one of these causes has no effect on
the possibility of failure for any of the other causes. Let us
assume that, if a cause $\underline{\Pi}^{(i)}$ were the only such cause of failure,

the lifetime would be equal to $\tau^{(i)}$. The question arises as to the lifetime τ under parallel action of all k causes. Obviously, from the instant at which one of the causes has already brought about a failure in the object, the action of other causes cannot change the situation. Therefore, the time during which the object will operate without failure is the time until the instant at which one of the causes $\underline{\Pi}^{(1)}, \underline{\Pi}^{(2)}, \ldots, \underline{\Pi}^{(k)}$ brings about a failure. If the ith cause has brought about a failure, then the lifetime is, in our notation $\tau^{(i)}$. Of course, if it is possible to remove the cause $\underline{\Pi}^{(i)}$, then the lifetime of the object will increase since there have been no failures up to instant $\tau^{(i)}$ due to the other causes. What we have said enables us to assert that the lifetime τ is given by the equation

$$\tau = \min(\tau^{(1)}, \tau^{(2)}, \ldots, \tau^{(k)}), \tag{190}$$

which means that this time is the minimum of the quantities $\tau^{(i)}$ for $i = 1, \ldots, k$. If $F_i(T)$ is the distribution function of the lifetime $\tau^{(i)}$ when only the ith cause is in effect, then the distribution function of the time τ can be represented

$$F(T) = 1 - [1 - F_1(T)][1 - F_2(T)] \ldots [1 - F_k(T)]. \tag{191}$$

Let us look at some examples. As our first example, let us consider the situation that exists when instantaneous injuries act in parallel that is, when there are failures due to "peak" loads and a scheme of accumulating injuries that leads to a normal distribution of the lifetimes. If $\underline{\Pi}^{(1)}$ is the cause consisting in wear at a constant rate and with strong mixing, then, in accordance with formulas (191), (33), and (103), we have

$$F(T) = 1 - e^{-\lambda T}\left[1 - \Phi\left(\frac{T - c}{\sigma}\right)\right]. \tag{192}$$

The density of this distribution is given by

$$f(T) = e^{-\lambda T}\left[\frac{1}{\sqrt{2\pi}\sigma}e^{-\frac{(T - c)^2}{2\sigma^2}} + \lambda\left(1 - \Phi\left(\frac{T - c}{\sigma}\right)\right)\right]. \tag{193}$$

The curves of the density (193) are shown in Fig. 56. If the mean
time $1/\lambda$ until a failure due to an instantaneous injury is less
than c (the mean time until the occurrence of a failure due to wear),
then the density curve resembles the curve of an exponential function
($\lambda = 2$). When $c < 1/\lambda$, the density has a well-defined hump and resembles a normal distribution.

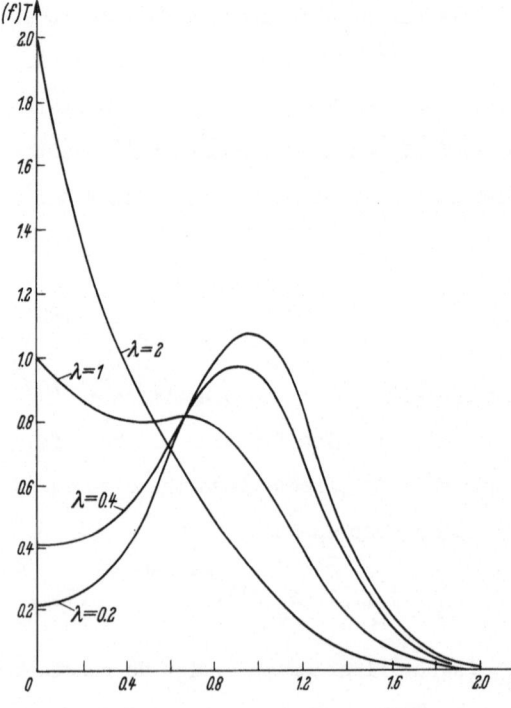

Fig. 56. The density
of the distribution
in accordance with
(193) for $c = 1, \sigma = 1/3$,
and various values of
the parameter λ.

The distribution function (192) is extensively used to describe
failures in vacuum tubes. Physically, this is explained by the fact
that failure of the tube is caused either by a parameter (for example
steepness of the characteristic) exceeding given limits or by an instantaneous breakdown in the tube (overheating, breaking of the tube,
etc.) under the influence of a "peak" in temperature or other influences. In connection with the extensive use of the distribution
(192), let us pause to look at the possibilities of estimating its
parameters.

If it is possible to separate the failures due to "peak" loadings
from failures due to wear, this still cannot serve as a basis for a

118

separate determination of the parameters of the distributions $F_1(T)$ and $F_2(T)$. To see this, we need only consider the following situation:

Suppose that the wear process proceeds very slowly. Therefore, if we consider its action separately, we need to observe the values of the lifetimes of large duration. Suppose also that the intensity of instantaneous failures is extremely great. Therefore, the lifetime corresponding to them, is, as a rule, small. Observing the action of the two causes simultaneously, we see basically rather small values of τ. Therefore, if we single out those failures resulting from wear, we shall get the false impression that the wear proceeds at a rapid rate and that the lifetime when it is the only cause of failure that is in effect is small.

Estimate of the parameters of a distribution. For an isolated estimate of the parameters of a distribution, we need to use the data on the sample functions of the wear. The method of finding the parameters from the sample functions of the wear was expounded in Chapter III. When we have found the parameters of the distribution $F_2(T)$, we can use the data on the quantities $\tau_1, \tau_2, \ldots, \tau_N$ to find the distribution parameter λ. To do this, we need only use the method of discrimination partitions. Since only the single parameter λ appears in the distribution $F_1(T)$, it follows that the set of quantities τ_i must be partitioned into two classes. We denote by Θ the boundary of the partition. If $\nu(\Theta)$ denotes the frequency of the values τ_i lying to the left of the boundary Θ, then, to find the parameter λ, we need to use the equation

$$e^{-\lambda\Theta}\left[1 - \Phi\left(\frac{\Theta - c}{\sigma}\right)\right] = 1 - \nu(\Theta),$$

from which we get

$$\lambda = \frac{\ln\left[1 - \Phi\left(\frac{\Theta - c}{\sigma}\right)\right] - \ln[1 - \nu(0)]}{\Theta}. \tag{194}$$

The quantity λ can be calculated if we know the parameters c and σ that appear in this formula.

To make clear the method of estimating the parameters on the basis of nothing more than the data on the lifetime, we need to note the following circumstance.

Let $\tau^{(1)}$ denote a random variable having an exponential distribution with parameter λ. Let $\tau^{(2)}$ denote a random variable that is independent of $\tau^{(1)}$ and that has a normal distribution with parameters c and σ. It follows from the preceding discussion that the random variable

$$\tau = \min(\tau^{(1)}, \tau^{(2)})$$

obeys the distribution (192).

Suppose that we have s independent sample functions $\tau_1, \tau_2, \ldots, \tau_s$ of a random variable τ. Let us assume that they are numbered in order of increasing size, that is, that $\tau_1 \leq \tau_2 \leq \ldots \leq \tau_s$. To each sample function τ_i are assigned the sample functions of the random variables $\tau^{(1)}$ and $\tau^{(2)}$. We denote them by $\tau_i^{(1)}$ and $\tau_i^{(2)}$, respectively. It follows form the definitions of τ and τ_1 that

$$\tau_1 = \min\{\min(\tau_1^{(1)}, \tau_1^{(2)}); \ldots; \min(\tau_s^{(1)}, \tau_s^{(2)})\} =$$

$$= \min\{\min(\tau_1^{(1)}, \ldots, \tau_s^{(1)}); \min(\tau_1^{(2)}, \ldots, \tau_s^{(2)})\} =$$

$$= \min(\tau_{min}^{(1)}, \tau_{min}^{(2)}),$$

where

$$\tau_{min}^{(1)} = \min(\tau_1^{(1)}, \ldots, \tau_s^{(1)}), \quad \tau_{min}^{(2)} = \min(\tau_1^{(2)}, \ldots, \tau_s^{(2)}).$$

Here, $\tau_{min}^{(1)}$ is the minimum of the s independent exponentially distributed quantities and $\tau_{min}^{(2)}$ is the minimum of the s independent normally distributed quantities.

It is easy to show that $\tau_{min}^{(1)}$ has an exponential distribution with parameter λs, that is, with mathematical expectation only $1/s$ that of $\tau^{(1)}$. One can also show that, for large s, the mathematical expectation of $\tau_{min}^{(2)}$ will be approximately $c - \sigma\sqrt{\ln s}$, that is, with increase in s, it will decrease extremely slowly [9]. Usually, $1/\lambda$

and c are quantities of approximately the same order of magnitude and, as a rule, $1/\lambda < c$. Then, the property of the mathematical expectations just mentioned leads to the fact that the random variable $\tau_{min}^{(1)}$ will itself influence the distribution of τ_1 in a definite way. Thus, for $1/\lambda = c$ and $s = 10$, the quantity τ_1 will have an almost exponential distribution.

This means that we may assume that τ_1 is, to all intents and purposes, distributed in accordance with an exponential law with parameter $\lambda^* = \lambda s$. This property actually is used in estimating the parameters. We recommend the following procedure.

1. The data on the lifetime $\tau_1, \ldots, \tau_N$ are subdivided into l groups with s values in each, so that $ls = N$. The choice of quantities τ_i must be made randomly, for example, by lots or by using a table of random numbers.

2. In each of the l groups, we seek the smallest value τ_n^*, for $n = 1, 2, \ldots, l$.

3. The parameter λ is estimated in accordance with the formula

$$\lambda = \frac{l}{s} \, \frac{1}{\displaystyle\sum_{n=1}^{l} \tau_n^*} \, . \tag{195}$$

Certain properties of such an estimate of the parameter of the exponential factor of distributions of the type (192) are considered in [41].

4. We choose partition points Θ_1 and Θ_2 with $\Theta_1 < \Theta_2$ and we calculate the number $m(\Theta_j)$ of values τ_i, for $i = 1, \ldots, N$, lying in the intervals $(0, \Theta_j)$ for $j = 1, 2$. We find the ratios $\nu(\Theta_j) = m(\Theta_j)/N$.

5. The parameters c and σ are estimated in accordance with the formulas

$$\left. \begin{aligned} c &- \frac{\Psi_1 \Theta_2 - \Theta_1 \Psi_2}{\Psi_1 - \Psi_2}, \\[2mm] \sigma &= \frac{\Theta_1 - \Theta_2}{\Psi_1 - \Psi_2}, \end{aligned} \right\} \tag{196}$$

where

$$\Psi_j = \Psi[1 - (1 - \nu(\Theta_j)) e^{\lambda \Theta_j}], \quad j = 1,2;$$

Ψ being the inverse Laplace function.

To estimate the parameters, we need to have at least 100-200 data and we need to choose s = 10-15.

The Weibull-Gnedenko distribution

Let us consider a system consisting of a group of elements and possessing the following properties: The failures of the elements are mutually independent. Failure of any element is treated as failure of the entire system. We call such systems chain systems.

Let τ denote the lifetime of the chain system and let $\tau^{(i)}$ denote the lifetime of the ith element of the system for $i = 1, 2, \ldots, k$. In such a case, as we have pointed out,

$$\tau = \min(\tau^{(1)}, \tau^{(2)}, \ldots, \tau^{(k)}).$$

We point out an important special case. Suppose that all elements of the chain system have an exponential distribution of the lifetime. In accordance with formulas (191) and (33), the distribution function F(T) is equal to

$$F(T) = 1 - e^{-\sum_{i=1}^{k} \lambda_i T}, \tag{197}$$

where the λ_i are parameters of the distributions of the elements of the chain system.

Of especial interest is the situation generalizing the case just described but having the following feature. The number k of elements of the chain system is great and all the distribution functions $F_i(T)$ are such that

$$F_i(T) = gT^{\gamma} + o(T^{\gamma}), \tag{198}$$

where g and γ are positive, as $T \to 0$.

122

This relationship determines the order of the infinitesimal F(T) for small T. In particular, equation (198) holds for a gamma distribution. One can show that, for large k, the distribution function F(T) is well-approximated by an expression of the form

$$F(T) = \begin{cases} 1 - \exp(-T^{\gamma}/\beta), & T \geq 0 \\ 0, & T < 0 \end{cases} \tag{199}$$

where

$$\beta = \frac{1}{gk}.$$

This distribution was proposed by W. Weibull [39] in 1939 without mathematical foundation. A rigorous mathematical treatment of related problems was done by B. V. Gnedenko in 1941 in the article [9].

In mathematical statistics, it is customary to call the distribution (199) a distribution of the third type for boundary terms of a sequence of independent variables. In what follows, we shall call (199) the Weibull- Gnedenko distribution. The density of this distribution has the form

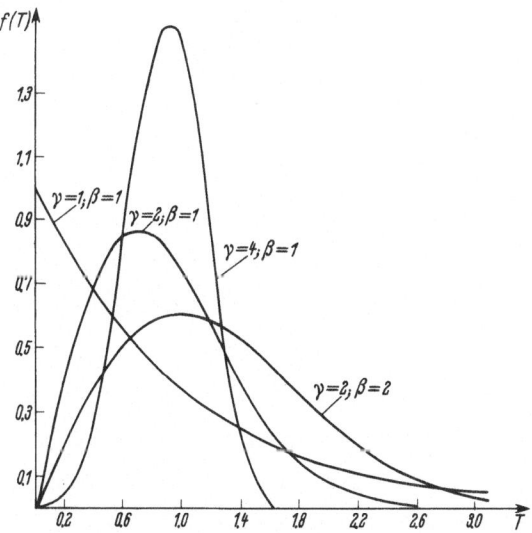

Fig. 57. The density of the Weibull-Gnedenko distribution for different values of the parameters γ and β.

$$f(T) = \begin{cases} \exp(-T^\gamma/\beta)\dfrac{\gamma}{\beta}\,T^{\gamma-1}, & T \geq 0 \\ 0 & , \ T < 0 \end{cases} \qquad (200)$$

The density curves are shown in Fig. 57. As we can see from the graphs, this distribution is asymmetric. In the special case of $\gamma = 1$, it is the exponential distribution (34).

The Weibull-Gnedenko distribution occupies an important position among the distributions of the lifetime. This is due to the fact that a small difference in the distributions of the lifetime of the elements does not hinder use of the distribution (199) to describe the lifetime of a device. Thus, if each of the elements has a gamma distribution of lifetime but the parameters of these distributions vary somewhat from element to element, then, for a sufficiently large number of elements, the distribution of the lifetime will be well-approximated by formula (199). To this we need to add that many devices contain a considerable number of elements that are identical or quite similar in their construction and that are used under approximately equal conditions. For example, an internal-combustion motor has several cylinders, a gas turbine has a large number of blades, a radio-receiver has a considerable number of condensers, resistors, and so forth. If the elements that are repeated in a single device are the determining factors for the lifetime of the devices, we have a scheme that leads to a Weibull-Gnedenko distribution.

It is shown in a number of works on reliability that a Weibull-Gnedenko distribution gives a good description of the distribution of the lifetime of numerous elements in radio-electric equipment-vacuum tubes, transistors, etc. — when the failure of these elements is regarded as the exceeding of established limits on the part of any of the parameters (test of the length of service). To a certain degree, this is in accordance with theoretical assumptions made in the derivation of the distribution (199). Thus, we may assume that the random changes in the parameters of these elements are weakly connected random processes. Then, if $\tau^{(i)}$ is the random length of service with respect to the ith parameter, the lifetime τ of the element is completely determined by the relationship given above: $\tau = \min(\tau^{(1)}, \ldots, \tau^{(k)})$.

Estimation of the parameters of the Weibull-Gnedenko distribution.

Suppose that we know the values $\tau_1, \ldots, \tau_N$ of the lifetime of N devices. To estimate the parameters γ and β, we use the method of discrimination partitions.

1. We choose the partition boundaries Θ_1 and Θ_2 with $\Theta_1 < \Theta_2$ and we calculate the numbers of values τ_i that lie in the ranges $(0, \Theta_j)$ for $j = 1,2$. Let us denote these numbers by $m(\Theta_j)$. We find the ratios $\nu(\Theta_j) = m(\Theta_j)/N$.

2. The parameters γ and β are estimated in accordance with the formulas

$$\left.\begin{aligned}
\gamma &= \frac{\ln\ln\left[\dfrac{1}{1 - \nu(\Theta_1)}\right] - \ln\ln\left[\dfrac{1}{1 - \nu(\Theta_2)}\right]}{\ln\Theta_1 - \ln\Theta_2}, \\[2em]
\beta &= \frac{\Theta_1^{\gamma}}{\ln\left[\dfrac{1}{1 - \nu(\Theta_1)}\right]}.
\end{aligned}\right\} \tag{201}$$

To get a crude estimate of the parameters of the distribution and also to check how well the experimental data can be smoothed off, it is convenient to use probability paper (see Fig. 58). The rules for its use are described in Example 8. To facilitate the calculations in accordance with formulas (201), we give the values of the functions

$$y_1(\chi) = \ln\frac{1}{1 - \chi}, \; y_2(\chi) = \ln\ln\frac{1}{1 - \chi}.$$

in Tab. 3 of the Appendix.

Example 8. Tab. 9 shows the result of testing of a lot of 100 special germanium transistors [23, p.53]. If any parameter of the transistor got outside established limits, a failure was assumed to have occurred. Let us choose $\Theta_1 = 200$ and $\Theta_2 = 2100$. In the present case, $\nu(\Theta_1) = 0.36$ and $\nu(\Theta_2) = 0.79$. From formulas (201), we find $\gamma = 0.532$ and $\beta = 37.6$.

Taking into account that the second column of Tab. 9 contains the

accumulated frequencies, let us plot them on probability paper
(see Fig. 58), and draw a straight line as nearly as we can through
the points obtained. We see that the points are located comparatively close to the straight line. Let us estimate the parameters γ
and β from Fig. 58. To do this, we use the special point A with
coordinates (2.718, 0.632), the vertical line H with abscissa 1,
and the special scale m. To estimate the value of γ, we need to
find the ordinate, with respect to the scale m, of the point B

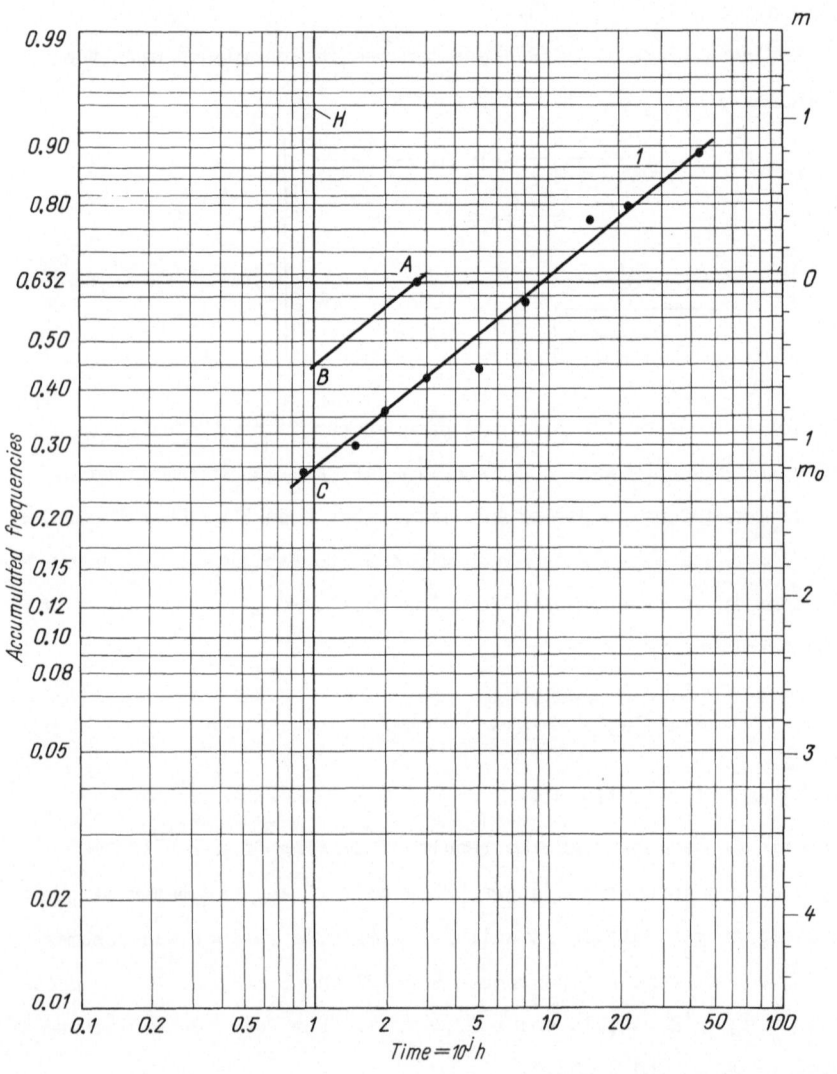

Fig. 58. Probability paper for the Weibull-Gnedenko distribution.

representing the intersection of the line H and the ray drawn
through A parallel to the line 1. In the present case, $\gamma = 0.52$.
To estimate β, we need to find the ordinate m_0, with respect to
the scale m, of the point C representing the intersection of the
line 1 with the line H.

Table 9

Results of testing of transistors

Testing time (in hours) from the instant the tests began	The portion of transistors that failed up to the instant of testing
90	0.26
150	0.30
200	0.36
300	0.42
500	0.44
800	0.57
1 500	0.77
2 100	0.79
4 300	0.87

The parameter β is found from the formula

$$\beta = e^{m_0} 10^{j\gamma}, \tag{202}$$

where j is the order of the time-scale factor (in the present case,
$j = 2$).

From Fig. 58 we find $m_0 = 1.16$; from equation (202), we find $\beta = 35.0$.
These estimates coincided satisfactorily with the estimates obtained
in accordance with the method of discrimination partitions.

We note that exponential paper (see Fig. 9) is a particular case
of the paper of Fig. 58 since an exponential distribution is a
special case of the distribution (199). Smoothing of the data obtained in accordance with an exponential distribution can also be
done on the probability paper of the Weibull-Gnedenko distribution.

In Chapters II and III, when we were examining the exponential and
gamma distributions, we described situations in which there is a

threshold of sensitivity t_0 that leads to a displacement of the distribution. If each of the elements belonging to a chain system has the same threshold of sensitivity t_0, then the system as a whole will also have this same threshold of sensitivity. Under the assumption that the approximate equation

$$\underline{P}\{\tau \le t_0 + T\} = F(t_0 + T) = gT^{\gamma} + o(T^{\gamma}),$$ (203)

which is analogous to equation (198), holds for small values of T, we arrive, for large k, at a Weibull-Gnedenko distribution with three parameters

$$F(T) = \begin{cases} 1 - \exp\left[- (T - t_0)^{\gamma}/\beta \right], & T \ge t_0, \\ 0 & , T < t_0. \end{cases}$$ (204)

The expression for the density has the form

$$f(T) = \begin{cases} \exp\left[- (T - t_0)^{\gamma}/\beta \right] \frac{\gamma}{\beta}(T - t_0)^{\gamma - 1}, & T \ge t_0 \\ 0 & , T < t_0 \end{cases}$$ (205)

Estimation of the parameters of a Weibull-Gnedenko distribution with threshold of sensitivity. Let $\tau_1, \ldots, \tau_N$ denote lifetimes of N devices.

1. We choose partition boundaries Θ_1, Θ_2, and Θ_3 such that $\Theta_1 < \Theta_2 < \Theta_3$ and calculate the number $m(\Theta_j)$ of values τ_i that lie in the intervals $(0, \Theta_j)$ for $j = 1, 2, 3$. We find the ratios $\nu(\Theta_j) = m(\Theta_j)/N$.

2. The coefficient z is determined in accordance with the formula

$$z = \frac{y_1 - y_2}{y_1 - y_3},$$

where

$$y_j = \ln \ln \frac{1}{1 - \nu(\Theta_j)}, \quad j = 1, 2, 3.$$

3. The parameter t_0 is found by solving the transcendental equation

$$(\Theta_2 - t_0) = (\Theta_3 - t_0)^z (\Theta_1 - t_0)^{1 - z}.$$ (206)

We seek a solution in the interval $0 < t_0 < \Theta_1$. One can show that this equation has a unique solution if

$$\frac{\Theta_2}{\Theta_1} < \left(\frac{\Theta_3}{\Theta_1}\right)^z.$$
(207)

Therefore, before solving the equation we need to test whether this condition is satisfied or not. If it is not satisfied, we need to change the partition points.

Equation (206) is easily solved with the aid of a graph.

4. The parameters γ and β are estimated in accordance with the formulas

$$\left.\begin{array}{l} \gamma = \dfrac{y_1 - y_2}{\ln(\Theta_1 - t_0) - \ln(\Theta_2 - t_0)}, \\[3ex] \beta = \dfrac{(\Theta_1 - t_0)^\gamma}{\ln\left[\dfrac{1}{1 - \nu(\Theta_1)}\right]}. \end{array}\right\}$$
(208)

To get a very crude estimate of the quantity t_0, we can use probability paper. The procedure for estimating t_0 consists in choosing a bias value t_0 of the initial data in such a way that $\tau'_i = \tau_i - t_0$ is sufficiently well-smoothed out in accordance with the distribution (199).

It is convenient to combine the computational method described above with the method of choosing t_0 on functional paper.

The Weibull-Gnedenko functional distribution

The initial relation (198) can be written in the more general form

$$F_i(T) = u^\gamma(T) + o(u^\gamma(T)),$$

where $u(T)$ is a monotonic function that approaches zero as $T \to 0$. The relation (198) is obtained if we set $u(T) = g^{1/\gamma}T$.

Let us consider the case in which $F_i(T)$ is a logarithmic gamma distribution. Let us establish the form of $u(T)$. Suppose that the mathematical expectation of the rate of wear is given by the equation

$$\underline{M}\{\xi(t)\} = \frac{1}{1+t}.$$ (209)

Then, the accordance with (125) and (126), the distribution function of the lifetime of an element is represented in the form

$$F_i(T) = \sum_{m=r}^{\infty} \frac{[\lambda \ln(1+T)]^m}{m!} e^{-\lambda \ln(1+T)}.$$ (210)

As $T \to 0$,

$$F_i(T) \approx \frac{[\lambda \ln(1+T)]^r}{r!}.$$

Thus, $u(T)$ is given by the equation

$$u(T) = C \ln(1+T),$$ (211)

where r plays the role of the exponent γ.

We shall not stop for a detailed proof but point out that, if γ is great, then the distribution function of the lifetime of a chain system is approximated satisfactorily by an expression of the form

$$F(T) = \begin{cases} 1 - \exp\{-[\ln(1+T)]^\gamma/\beta\}, & T \geq 0, \\ 0, & T < 0. \end{cases}$$ (212)

This expression is a particular case of the Weibull-Gnedenko functional distribution defined by the formula

$$F(T) = \begin{cases} 1 - \exp\{-[u(T)]^\gamma/\beta\}, & T \geq 0, \\ 0, & T < 0. \end{cases}$$ (213)

In works on fatigue longevity, one uses the distribution (212) with displacement parameter. A very similar distribution is used in Weibull's article [40]. More precisely, the distribution he used was of the form

$$F(T) = \begin{cases} 1 - e^{-\frac{[\log(T - t_0 + 1)]^\gamma}{\beta}}, & T \geq t_0, \\ 0, & T < t_0, \end{cases}$$

where β, γ, and t_0 are the parameters of the distribution.

We estimate the parameters of this distribution by the same methods as for the distribution (204). The only difference is that we need to shift from the quantity $\tau_1, \ldots, \tau_N$ to the quantities $u_1 = \log \tau_1, \ldots, u_N = \log \tau_N$.

Chapter VI

The Failure Rate

The failure rate and the residual lifetime

In our everyday life, we are frequently much more interested in
the reliability of used goods than new ones. When someone buys a
clock, the consumer has a warranty from the store ensuring that
he can replace the clock with a new one or have it repaired if the
clock fails within a certain period of time. Therefore, the ability
of the clock to function without failure during this period of time
does not trouble the user greatly. The question as to how long the
clock will run $\underline{\text{after}}$ the period specified in the warranty is quite
another matter. Failure of the clock then lies entirely on the
shoulders of the purchaser. What the purchaser is interested in is
that the probability that the clock will run without failure over
a period of several years after the period specified in the warranty is sufficiently great. Let us look at this situation from a
formal point of view. Let us denote by t_H a period of time during
which the object in question functioned without failure and let us
introduce the probability $\underline{P}\{\tau_k > T | t_H\}$ of failure-free operation
during a period of time T after the object has functioned without
failure for a time t_H. Thus, in the case of our present example,
if the period of the warranty is a year, then $t_H = 1$. Wishing to
determine the probability that the clock will continue to function
for 20 years after the expiration of the warranty, we need to cal-
culate the probability $\underline{P}\{\tau_h > 20 | t_H = 1\}$. This probability is a con-
ditional probability and, to evaluate it, we need to make use of
the fact that an arbitrary distribution of the lifetime can be
formally written

$$F(T) = 1 - \exp\left[-\int_0^T \lambda(t)dt \right] \qquad (214)$$

where $\lambda(t)$ denotes the function

$$\lambda(t) = \frac{f(t)}{1 - F(t)} \, . \tag{215}$$

One can easily verify directly the validity of (214). The function $\lambda(t)$ is used extensively in reliability theory and known as the failure rate. Its physical meaning will be clarified below.

The conditional probability $\underline{P}\{\tau_k > T \mid t_H\}$ is calculated in accordance with the general rules as the ratio of the probability $\underline{P}\{\tau > T + t_H\}$ that there were no failures during the period $t_H + T$ to the probability $\underline{P}\{\tau > t_H\}$ that there were no failures in the period of time t_H. Beginning with formula (214), we can write this ratio as follows:

$$\underline{P}\{\tau_k > T \mid t_H\} = \frac{\underline{P}\{\tau > T + t_H\}}{\underline{P}\{\tau > t_H\}} =$$

$$= \exp\left[- \int_0^{t_H + T} \lambda(t)\,dt \right] \Big/ \exp\left[- \int_0^{t_H} \lambda(t)\,dt \right]$$

and, consequently,

$$\underline{P}\{\tau_k > T \mid t_H\} = \exp\left[- \int_{t_H}^{t_H + T} \lambda(t)\,dt \right]. \tag{216}$$

Suppose that the quantity T is small. We denote it by Δ_T. Then, formula (216) can be represented in the form

$$\underline{P}\{\tau_k > \Delta_T \mid t_H\} \approx e^{-\lambda(t_H)\Delta_T}. \tag{217}$$

The probability $\underline{P}\{\tau_k > \Delta_T \mid t_H\}$ is the probability that the object will continue to function a time Δ_T after it has functioned without failure for a time t_H. Thus, formula (217) enables us, as it were, to extrapolate the possibility of failure-free operation of the object for a small period of time Δ_T in advance if the object is observed at the instant t_H during its failure-free operation. It follows from formula (217) that the greater the value of $\lambda(t_H)$ the less probable it will be that the object will continue to operate for the additional time Δ_T. Thus, the quantity $\lambda(t_H)$ provides the failure rate for a period of time close to t_H. For just this reason, the time function $\lambda(t)$ is called either the failure rate or the "risk of failure"

Of course, the owner of the clock would be pleased to have the failure rate $\lambda(t_H)$ decrease with increasing time t_H. Then, the "risk of failure" would always decrease and after the expiration of the time specified in the warranty and he could go for a comparatively long while without having the clock repaired.

In order to clarify the meaning of the failure rate $\lambda(t)$ in greater detail, let us consider it from several other points of view.

Let us look again at the formula defining $\lambda(t)$. Let us represent this formula in the form

$$\lambda(t)\Delta_t = \frac{f(t)\Delta_t N}{[1 - F(t)]N},$$

where N is the number of exemplaires of the object in question that are being observed.

The product $f(t)\Delta_t$ is the probability of failure of the object during the interval from t to $t + \Delta_t$. Correspondingly, $f(t)\Delta_t N$ is the average number of objects that fail during the time from t to $t + \Delta_t$. The product $[1 - F(t)]N$ is the average number of objects that do not fail during the time t. Thus, the product $\lambda(t)\Delta_t$ is the ratio of the number of objects that fail during the time from t to $t + \Delta_t$ to the number of objects that remain able to function up to the instant t.

The procedure for constructing graphs of the failure rate from data and the lifetime is described in detail in texts on reliability theory [25, 34].

The convenience of the concept of failure rate from a standpoint of estimating it from experimental data consists in the fact that the quantities $\lambda(t)\Delta_t$ can be calculated from the course of the tests without waiting until all the objects tested fail. However, as a consequence of the fact that $\lambda(t)\Delta_t$ is actually a frequency, it is subject to large random variations, especially near the end of the tests when the number of objects that have not failed is small. Therefore, to get a reliable estimate of the failure rate, it is necessary to have a large number of objects: 100 or more.

Let us return to formula (217) and let us show that the quantity $\lambda(t_H)\Delta_t$ is simply the conditional probability of failure during the interval $(t_H, t_H + \Delta_t)$ under the assumption that the object has operated without failure for a time t_H. Since

$$\underline{P}\{\tau_k \leq \Delta_t | t_H\} = 1 - \underline{P}\{\tau_k > \Delta_t | t_H\},$$

we have

$$\underline{P}\{\tau_k \leq \Delta_t | t_H\} \approx 1 - e^{-\lambda(t_H)\Delta_t}.$$

If we expand the right-hand member in a series of powers of Δ_t, we obtain

$$\underline{P}\{\tau_k \leq \Delta_t | t_H\} = \lambda(t_H)\Delta_t + o(\Delta_t). \tag{218}$$

The failure rate $\lambda(t)$ is closely connected with the conditional mathematical expectation of the remaining lifetime computed under the assumption that the object did not fail in the interval $(0, t_H)$. We shall refer to the random variable representing the lifetime calculated from some fixed instant of time t_H until the instant of failure under the condition that there was no failure prior to the instant t_H as the residual lifetime and we shall denote it by $\tau_0(t_H)$.

If we have data $\tau_1, \ldots, \tau_N$ on the lifetime, then an empirical estimate of the mathematical expectation $\underline{M}\{\tau_0(t_H)\}$ of the residual time is obtained as follows: We calculate the quantities

$$\tau_0^{(j)}(t_H) = \tau_j - t_H$$

for those values of τ_j that are no less than t_H (the residual time is a nonnegative quantity). Suppose that there are L such quantities. Then (with small bias),

$$\underline{M}\{\tau_0(t_H)\} \approx \overline{\tau}_0(t_H) = \frac{1}{L}\sum_{k=1}^{L} \tau_0^{(j_k)}(t_H). \tag{219}$$

In Fig. 59a, one can see that for $t_H > 40$ hr there is a systematic bias in the estimate for $\overline{\tau}_0(t_H)$ (the curve for the mean residual time is systematically distributed lower than the quantity $\overline{\tau} = \tau_0(0)$). An improved estimate for the quantity $M\{\tau_0(t_H)\}$ can be obtained by using the method of maximum likelihood [22]. The problem of finding an estimate $\bar{\tau}_0(t_H)$ for the quantities $M\{\tau_0(t_H)\}$ reduces to solving the equation

$$\bar{\tau}_0(t_H) + (\frac{N}{L} - 1)t_H \frac{1}{e^{t_H/\bar{\tau}_0(t_H)} - 1} = \frac{1}{L} \sum_{k=1}^{L} \tau_0^{j_k}(t_H) + t_H. \quad (219a)$$

This equation has a unique positive root for all $L \geq 1$ and $t_H \geq 0$. For $t_H = 0$, the estimate $\bar{\tau}_0(t_H)$ coincides with the usual estimate of the parameter $1/\lambda$ of the exponential distribution.

The distribution function of the residual time is given, in accordance with (216), by the equation

$$\underline{P}\{\tau_0(t_H) \leq T\} = F(T \mid t_H) = 1 - \exp\left[- \int_{t_H}^{t_H+T} \lambda(t)dt \right]. \quad (220)$$

From this we can easily obtain an expression for the mathematical expectation of the residual time (average residual time):

$$\underline{M}\{\tau_0(t_H)\} = \int_0^\infty \exp\left[- \int_{t_H}^{t_H+T} \lambda(t)dt \right]dT. \quad (221)$$

This formuala gives the relationship between the average residual time and the failure rate. From it, we derive, in particular, the following differential equation:

$$\frac{d\underline{M}\{\tau_0(t_H)\}}{dt_H} = \lambda(t_H)\underline{M}\{\tau_0(t_H)\} - 1, \quad (222)$$

which enables us to calculate $\lambda(t_H)$ if we know the expression for the residual mathematical expectations $\underline{M}\{\tau_0(t_H)\}$.

Study of the behavior of $\bar{\tau}_0(t_H)$ from the experimental data has the defect that it is possible only when we have finished testing all the objects. However, an estimate of the residual average times has considerably fewer random fluctuations than the estimate of the failure rate calculated from the same data. This is explained by the better statistical properties of estimates of the mathematical expectation in comparison with estimates of frequencies. This is shown graphically in Fig. 59. Fig. 59a shows a graph of the behavior of the residual averages for the data of Tab. 1. (As a preliminary, the data of the table were displaced to the left by an amount equal to the value t_0 of the threshold of sensitivity $t_0 =$ 36 h). The solid curve shows the residual means calculated in accordance with formula (219). The dashed curve shows the residual means obtained by solving equation (219a). Fig. 59b shows the graph of the behavior of the failure rate that was constructed from the same data. In this case, the deviations in the failure rates from the theoretical average level are considerably greater.

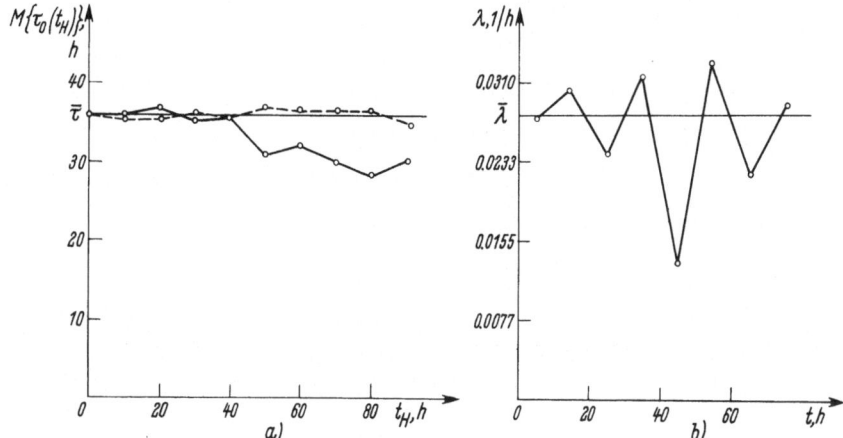

Fig. 59. The behavior of the average residual time (a) and the failure rate (b) from the data of Tab. 1. $\bar{\lambda}$ and $\bar{\tau}$ are the corresponding theoretical values.

Let us consider separately the case of an exponential distribution. Its function has the form

$$F(T) = 1 - e^{-\lambda T}.$$

One can easily see that, for this distribution, the failure rate is constant and equal to

$$\lambda(T) = \lambda.$$

From (221), we can show that

$$\underline{M}\{\tau_0(t_H)\} = \frac{1}{\lambda}.$$

Thus, for an exponential distribution, the average residual time coincides with the average time of a "new" object. In this sense,

objects obeying an exponential law that have operated for a period
of time without failure are no worse than new ones. It follows
from this, in particular, that there is no sense in automatically
replacing such objects under preventive maintenance.

Monotonically increasing failure rate

For a condiderable number of objects and systems, a constant increase in the failure rate $\lambda(t)$ is typical. A natural cause of
increase in the failure rate may be wear. When the wear is small,
the failure rate is comparatively small, but when the wear is
great, it becomes perceptible.

Let us look at the failure rate of a gamma distribution. The function of a gamma distribution is given by formula (73) and, on the
basis of equation (215), we can easily show that

$$\lambda(T) = \frac{\lambda^r T^{r-1}}{(r-1)! \left[1 + \frac{1}{1!} \lambda T + \frac{1}{2!} (\lambda T)^2 + \ldots + \frac{(\lambda T)^{r-1}}{(r-1)!} \right]}. \tag{223}$$

As T increases, $\lambda(T)$ increases monotonically and approaches a limit
λ as $T \to \infty$ (see Fig. 60).

A normal distribution also has a monotonically increasing failure
rate but, in contrast with the gamma distribution, the failure
rate $\lambda(T) \to \infty$ as $T \to \infty$ (see Fig. 61). An unbounded increase in the
failure rate is also observed for a Weibull-Gnedenko distribution
given by formula (199) when the parameter $\gamma > 1$. The expression for
the failure rate of this distribution has the simple form

$$\lambda(T) = \gamma \frac{T^{\gamma-1}}{\beta}. \tag{224}$$

Graphically, changes in the rate $\lambda(T)$ of the Weilbull-Gnedenko
distribution for various values of the parameter γ are shown in
Fig. 62. The existence of a threshold of sensitivity t_0 displaces
the rate curve to the right by an amount equal to t_0.

The distribution laws for the lifetime that has an increasing
failure rate are sometimes called laws of the "aging" type. A re-

sumé of the formulas for $F(T)$, $f(T)$, and $\lambda(T)$ is given in Tab. 10. We point out an interesting feature of the failure rate of the distribution (191). The density of this distribution has the form

$$f(T) = f_1(T) \prod_{i=2}^{k} (1 - F_i(T)) + \ldots + f_k(T) \prod_{i=1}^{k-1} (1 - F_i(T)). \qquad (225)$$

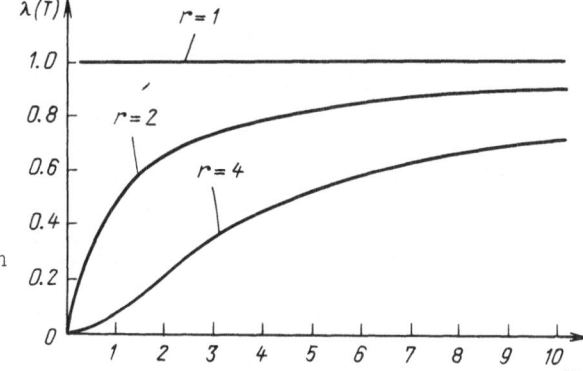

Fig. 60. Failure rate of gamma distribution with $\lambda = 1$.

From this it follows that the failure rate has the form

$$\lambda(T) = \frac{f(T)}{1 - F(T)} = \frac{f_1(T)}{1 - F_1(T)} + \ldots + \frac{f_k(T)}{1 - F_k(T)}, \qquad (226)$$

that is, it is equal to the sum of the failure rates of the distributions $F_i(T)$.

Thus, we have established the following fact: If the lifetime of

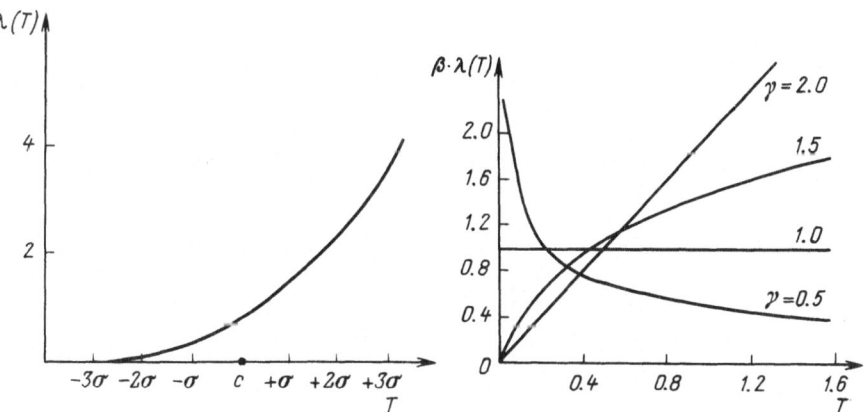

Fig. 61. Failure rate of a normal distribution.

Fig. 62. Failure rate of a Wei-bull-Gnedenko distribution.

Table 10. Resumé Table of Distributions.

Name of the distribution	Density $f(T)$	Distribution function $F(T)$	Failure rate $\lambda(T)$	Mathematical expectation $\underline{M}[\tau]$	Variance $\underline{D}[\tau]$
Exponential distribution *	$\lambda e^{-\lambda T}$	$1 - e^{-\lambda T}$	λ	$\dfrac{1}{\lambda}$	$\dfrac{1}{\lambda^2}$
Gamma distribution *	$\dfrac{\lambda_T^{r-1} e^{-\lambda T}}{\Gamma(r)}$ (r - an integer)	$1 - e^{-\lambda T} \sum\limits_{k=0}^{r-1} \dfrac{(\lambda T)^k}{k!}$	$\dfrac{\lambda^r T^{r-1}}{\sum\limits_{k=0}^{r-1} (r-1)!\,(\lambda T)^k}$	$\dfrac{r}{\lambda}$	$\dfrac{r}{\lambda^2}$
Normal distribution	$\dfrac{1}{\sqrt{2\pi}\sigma} \exp\left[-\dfrac{(T-c)^2}{2\sigma^2}\right]$	$\Phi\left(\dfrac{T-c}{\sigma}\right)$	$\dfrac{\exp\left[-\dfrac{(T-c)^2}{2\sigma^2}\right]}{\sqrt{2\pi}\,\sigma\,\Phi\left(\dfrac{c-T}{\sigma}\right)}$	c	σ^2
Logarithmic normal distribution *	$\dfrac{A}{\sqrt{2\pi}\sigma}\dfrac{1}{T} \exp\left[-\dfrac{(\log T - c)^2}{2\sigma^2}\right]$ $A = \log e = 0.4343$	$\Phi\left(\dfrac{\log T - c}{\sigma}\right)$	$\dfrac{A \exp\left[-\dfrac{(\log T - c)^2}{2\sigma^2}\right]}{\sqrt{2\pi}\,\sigma T\,\Phi\left(\dfrac{c-\log T}{\sigma}\right)}$	$\exp\left(\dfrac{c}{A} + \dfrac{\sigma^2}{2A^2}\right)$	$\exp\left[2\dfrac{c}{A} + \dfrac{\sigma^2}{A^2}\right] \times$ $\times\, [\exp(\sigma^2/A^2) - 1]$

Distribution	$f(T)$	$F(T)$	$\lambda(T)$		
Relaxational distribution	$\{\mu_C + \lambda[1 - \exp(-\mu T)]\} \times$ $\exp\{\frac{\lambda}{\mu}[1 - \exp(-\mu T)] - (\lambda + \mu_0)T\}$	$1 - \exp[\frac{\lambda}{\mu} - (\lambda + \mu_0)T - \frac{\lambda}{\mu}\exp(-\mu T)]$	$\mu_0 + \lambda(1 - e^{-\mu T})$	**	**
Weibull-Gnedenko distribution *	$e^{(-T^\gamma/\beta)}\,\frac{\gamma}{\beta}\,T^{\gamma-1}$	$1 - e^{(-T^\gamma/\beta)}$	$\frac{\gamma}{\beta}\,T^{\gamma-1}$	**	**
Distribution of the minimum of an exponentially distributed and of a normally distributed random variable	$\exp(-\lambda T)\left[\lambda\Phi\left(\frac{c-T}{\sigma}\right) + \frac{1}{\sigma}\frac{1}{\sqrt{2\pi}}\exp\left(-\frac{(T-c)^2}{2\sigma^2}\right)\right]$	$1 - e^{-\lambda T}\Phi\left(\frac{c-T}{\sigma}\right)$	$\lambda + \dfrac{\exp\left[-(T-c)^2/2\sigma^2\right]}{\sigma\sqrt{2\pi}\,\Phi[(c-T)/\sigma]}$	**	**

* When there is a threshold of sensitivity t_0 in the expressions for $f(T)$, $F(T)$, and $\lambda(T)$, instead of the argument T we have the difference $T - t_0$. The corresponding densities, distribution functions, and failure rates are equal to zero for $T < t_0$. To the mathematical expectations of the distributions is added the value of t_0. The variances remain unchanged by the presence of a threshold of sensitivity.

** These cannot be expressed in terms of elementary functions.

a system is determined as the minimum of a system of k independent
random variables having the distribution law $F_i(T)$ with failure
rate $\lambda_i(T)$, then the failure rate of the system is

$$\lambda(T) = \sum_{i=1}^{k} \lambda_i(T).$$

From this it follows, in particular, that the distribution of the
minimum of the system of random variables with laws of the "aging"
type is also a distribution of the "aging" type.

Nonmonotonic failure rate

In contemporary industry, special measures are taken to prevent
technological defects. In recent years, the effectiveness of these
measures has considerably increased as a result of the use of pres-
ent-day methods of nondestructive control: the use of X-rays, ul-
trasonic defect detectors, the use of colors to detect defects,
etc. Nonetheless, we need to remember that isolated random "flaws"
in a technological process will lead to the occurrence of weak
spots that will break down under peak loads. For example, we mention two parts held together by soldering. At the present time,
there are no sufficiently convenient industrial methods of control
of conductor connections that are made by soldering. Therefore, a
"weak" connection can remain unnoticed at the output control of a
plant, and this leads to its breakdown when it is used. To keep defective items from reaching the consumer unnoticed, manufacturers
resort to some "training"; that is, the items produced are subjected to a relatively short-time special testing during which the
load is close to the maximum load that is likely to be encountered
in use. Thus, manufactured objects that are likely to be subjected
to vibrational loads when they are used are tested on vibrating
stands. Tests of this nature are carried out in the hope that
"weak" points in defective items will show up and only items without technological defects will go to the consumer.
Tests of this nature cannot be too protracted since this would
lead to wear of the items in question and to a decrease in their

operating capacity. However, the time of testing cannot be too short either since the "weak" points in defective items may not show up immediately. The expense due to testing and the fear of wear on the items result in a certain, usually small, number of objects with defects passing on to the consumer.

Let us now look at the situation that we have described from a standpoint of computation. If a certain portion of the items have technological defects, weak points, and the basic portion is taken out of operation because of wear (aging) of the units, then the distribution of the lifetime is a superposition of two distributions: an exponential distribution for the defective items and a distribution corresponding to the nature of the wear for the others. We denote by $F_1(T)$ the distribution function of the lifetime of the defective items and we denote by ε the portion of these items in the total set. We denote by $F_2(T)$ the distribution function of the lifetime of the remaining items. Of course, their portion in the total set will be equal to $1 - \varepsilon$. Then, the distribution function for the entire set is represented as follows:

$$F(T) = \varepsilon F_1(T) + (1 - \varepsilon) F_2(T). \tag{227}$$

Let us stop to look at a frequently encountered situation when $F_2(T)$ is a gamma distribution. The density for (227) has, in accordance with (33) and (73), the form

$$f(T) = \varepsilon \lambda_1 e^{-\lambda_1 T} + (1 - \varepsilon) \frac{\lambda_2^r T^{r-1} e^{-\lambda_2 T}}{\Gamma(r)}. \tag{228}$$

Fig. 63 shows the graph of $f(T)$ for values of the parameters $\lambda_1 = \lambda_2 = 1$, $r = 5$, $\varepsilon = 0.1$ (curve 3) and separately the two components of the distribution (curves 1 and 2).

As a result of use for a period of time t_N, a considerable number of defective items will fail. The "surviving" items will, on the average, have higher longevity. Consequently, if an item that has been in operation for a time t_H is still in operating condition, the probability that it will fail during the interval from t_H to $t_H + \Delta_T$ will be less than the probability of its failure in the interval from 0 to Δ_T. For just this reason, we talk of training. The suitability of objects for training is expressed in the behavior of the failure rate $\lambda(T)$. Fig. 64 shows the failure rate for the distribution shown in Fig. 63. As one can see from Figure 64, there

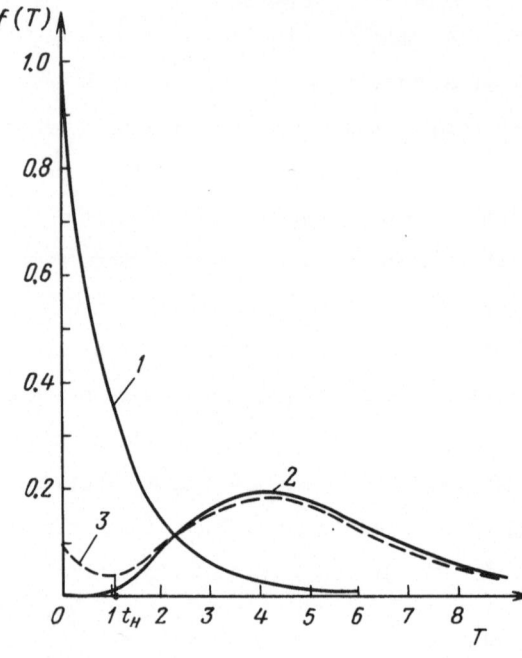

Fig. 63. The density of an exponential distribution (curve 1), a gamma distribution (curve 2), and their superposition with weights $\varepsilon = 0.1$ and $1 - \varepsilon = 0.9$ (curve 3.)

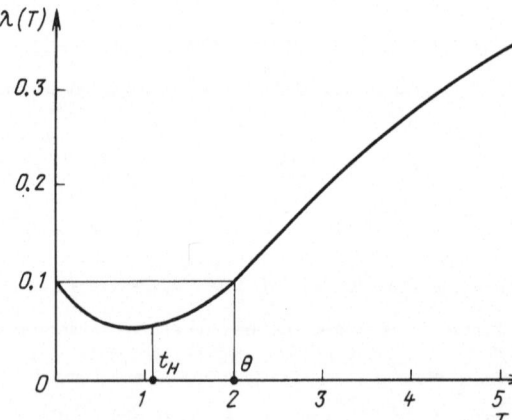

Fig. 64. The failure rate under superposition of an exponential distribution and a gamma distribution.

is a period of lowering of the failure rate. This period corresponds to the period of death of "weak" items or, in other words, the "period" of "natural selection". In what follows, as the wear of the elements increases, the failure rate will begin to rise and can attain large values.

144

It is obvious from Fig. 64 that, at least prior to the instant Θ, there is no sense in making a preventive maintenance of units since such a change could only lead to an increase in the failure rate. One can also see that if training were carried out under production conditions, it should be carried out up to an instant approximately equal to t_H when the main bulk of defective items has not yet failed. We note that, from the point of view of ensuring high reliability of trained objects, crude technological defects are less dangerous. Such defects show up easily in the training process. The situation is quite different when a technological defect is of a hidden nature, that is, the mean time before occurrence of failure of a defective item is comparable in magnitude with the mean time until failure as a result of wear of high-quality items. Fig. 64 shows a definite dip in the curve of $\lambda(T)$. Here, $\underline{M}\{\tau^{(1)}\} = 1/\lambda_1 = 1$ and $\underline{M}\{\tau^{(2)}\} = r/\lambda_2 = 5$: that is, the mean time until failure of a defective item is only one-fifth as great as the time until failure due to wear of the main portion of the items. If, instead, we had, for example, $r = 5$, $\lambda_1 = 1$, $\lambda_2 = 2.5$, and $\underline{M}\{\tau^{(2)}\}/\underline{M}\{\tau^{(1)}\} = 2$, the training would have no practical effect. This is true because, if the training were carried out until the failure of all "weak" items, this would lead to a significant amount of wear of all the other items. If the training were carried out for a short time, only a small percentage of the defective items would fail. All this means that, under production conditions, we need to watch out not so much for gross blunders as relatively minor violations in the course of the technological process. The maintenance of technological discipline and overall order is the deciding influence here.

Let us now return to equation (221). Suppose that $\lambda(t)$ increases monotonically with respect to t. Then, if $t_H^{(1)} > t_H^{(2)}$, it follows that

$$\underline{M}\{\tau_0(t_H^{(1)})\} > \underline{M}\{\tau_0(t_H^{(2)})\}. \tag{229}$$

This means that, for distribution laws of the aging type with increasing failure rate, the mathematical expectation of the residual time decreases monotonically with respect to t_H. This state of affairs in perfectly natural since, with increase in the preliminary time on test, t_H, the wear increases and the remainder of the longevity must decrease.

However, we can have an opposite type of situation, which at first glance seems strange. Specifically, we can have increase in the value of the mean residual time. Suppose that the failure rate $\lambda(t)$ begins to decrease monotonically at some point $t = t_H^*$. Then, if $t_H^* < t_H^{(1)} < t_H^{(2)}$, it follows from (221) that

$$\underline{M}\{\tau_0(t_H^{(1)})\} < \underline{M}\{\tau_0(t_H^{(2)})\}, \tag{230}$$

that is, the mean residual time will increase from some instant on. Furthermore, one can show, that $\underline{M}\{\tau_0(t_H)\} \to \infty$ as $\lambda(t) \to 0$ and, on the other hand, $\underline{M}\{\tau_0(t_H)\} \to 0$ as $\lambda(t) \to \infty$.

Let us look at some examples. Bernstein's distribution (152), which corresponds to a scheme of linear wear, has failure rate expressed by the formula

$$\lambda(T) = \frac{e^{-\dfrac{(T-c)^2}{2(aT^2+b)}}}{\sqrt{2\pi}\left[1 - \Phi\left(\dfrac{T-c}{\sqrt{aT^2+b}}\right)\right]} \; \frac{b+caT}{[b+aT^2]^{3/2}}. \tag{231}$$

One can easily see that $\lambda(T) \to 0$ as $T \to \infty$. The behavior of $\lambda(T)$ is shown in Fig. 65. The drop in failure rate after the instant Θ is explained by the fact that items possessing a high rate of wear can fail and the rate of wear of the "surviving" items is relatively small so that they have a greater longevity.

The failure rate of a logarithmic normal distribution given by formula (131) is of roughly the same nature:

$$\lambda(T) = \frac{A \exp\left[-(\log T - c)^2/2\sigma^2\right]}{\sqrt{2\pi}\,\sigma T[1 - \Phi(\log T - c)/\sigma]}. \tag{232}$$

One can see that $\lambda(T) \to 0$ as $T \to \infty$ in this case also. For both these distributions, there exists a point Θ of maximum rate. Therefore, we may assert that, from this instant on, the mean residual lifetime will increase.

For a Weibull-Gnedenko distribution when the parameter $\gamma < 1$, the failure rate decreases monotonically with respect to T:

$$\lambda(T) = \frac{\gamma}{\beta} T^{\gamma-1}. \tag{233}$$

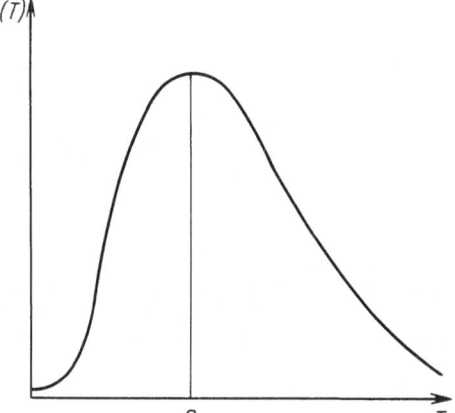

Fig. 65. Behavior of the failure rate of a logarithmic normal distribution and Bernstein's distribution.

In this case, $\underline{M}\{\tau_0(t_H)\}$ increases monotonically with respect to t_H on the entire interval $(0, \infty)$. Thus, for example, for $\gamma = 1/2$, we can obtain the expression (with $\beta = 1$)

$$\underline{M}\{\tau_0(t_H)\} = 2(1 + \sqrt{t_H}).$$

Thus, the nonmonotonic nature of the change in $\lambda(T)$, increase in the mean residual time, and the possibility of training are observed rather frequently.

Up to now, when we have been speaking of training, we referred to the failure of "weak" items. However, in a number of systems, we observe the property of "trainability" that cannot be explained on this basis. We cite, for example, the trainability of metals by means of the application of a small cyclical load. We know [27] that metals that are trained by a small cyclic load are able to withstand high cyclic loads in the course of a much more extended period of time than untrained ones. Here, the characteristic feature is the fact that we do not observe the failure of items in the training process. Therefore, we must assume that the training process entails a "strengthening of the health" of the objects. In the

147

theory of fatigue longevity of metals, the process of increasing
the durability of a metal in its training has been given the name
"strengthening".

On the basis of what has been said, we need to distinguish between
two types of training. The first type may be referred to as "burn-
in training". The term "burn-in" has source in the radio industry,
where "weak" items are "burned-in" by training under factory conditions. The second type of training is "training by strengthening".
Here, strengthening appears as a process increasing the resistance
to wear process.

To make clearer the concept of strengthening, we resort to an analogy with a process of instruction.[1] Let us look at the following
process of instruction of a rat in a labyrinth:[2]

Suppose that we have the labyrinth shown in Fig. 66. A rat is put
into the labyrinth. If the rat turns to the right, he is fed a
piece of meat: if he turns to the left he is fed poison. It is
assumed that if the rat takes r doses of poison he dies. The effect
of the instruction is that, if the rat turns to the right, then
the next time he is put into the labyrinth, the probability of his
turning to the right will increase. Thus, the more times the rat
turns to the right, the greater will be the probability that he
will continue to turn to the right and the less likely will he take
the next dose of poison. A rat that has learned to turn to the
right will have a high longevity, since if he also takes doses of
poison, there will long intervals of time between these doses. To
a considerable degree, the longevity of the rat is determined by
the course of the initial period of instruction. When he is put
into the labyrinth for the first time, still uninstructed, the probability of his turning to the left will be high (in instruction
theory, it is assumed that this probability is 0.5). If the instruction process turns out unsuccessful, the rat will obtain r

---

[1] This analogy was suggested to the authors by Academician Yu. V.
   L i n n i k .
[2] Models of instruction of this type are considered in the book
   [7].

doses of poison after a small number of trials. On the other hand,
if the rat makes many turns to the right in the initial period, the
effect of the instruction will show up afterwards and his longevity
will be high.

The process of instruction of a rat in a labyrinth can, to a certain
degree, be regarded as an analogue of the process of cyclic loading
of a metal. Each admission of the rat into the labyrinth can be regarded as a loading cycle. Successive doses of poison occur as isolated injuries and the effect of the instruction can be regarded as

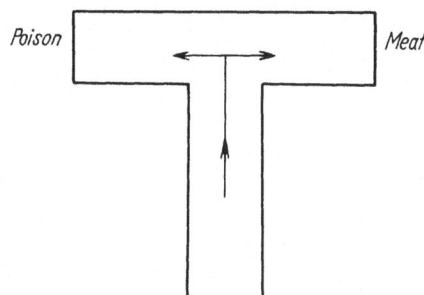

Fig. 66. The scheme of instruction of a rat in a labyrinth.

a strengthening process. An extremely important feature of this
scheme is the parallel action of the processes of injury and
strengthening. In metals, the strengthening factor is the smoothing
of the properties of their grains as a result of the redistribution
of the stresses. An injury consists in an accumulation of cavities
(the development of slip lines into cracks).

It must be emphasized that the training done here is unrelated to
the failure of weak items. Furthermore, during the initial period,
all the items are of the same quality. Variations in the lifetime
τ are determined exclusively by the randomness of the process of
accumulation of injuries. The training occurs as a sort of property
of "accommodation" to the loading conditions.

Let us turn now to a scheme of formation of a logarithmic gamma
distribution (see p. 74). In accordance with this scheme, the probability of injury during the time from T to $t + \Delta_T$ is given by

$$\gamma(T) = \frac{a\lambda}{1 + T} \Delta_T + o(\Delta_T).$$

Thus, the probability $\gamma(T)$ decreases with increasing time T. Decrease in the probability $\gamma(T)$ is a reflection of the strengthening process. The process of accumulation of injuries is a process of wear. Therefore, a scheme of formation of a logarithmic gamma distribution possesses by its nature the property of trainability of objects. An object that receives a small number of injuries over a long period t_H is strengthened and its longevity will be high. On the other hand, if an object receives a large number of injuries over a small period t_H, then, since the strengthening effect is as yet slight, it will have low longevity. What we have been saying shows up in the behavior of the failure rate the graph of which is similar to the one shown in Fig. 65.

Since a logarithmic normal distribution and a logarithmic Weibull-Gnedenko distribution are obtained as consequences of a logarithmic gamma distribution, they also reflect the presence of strengthening processes. Therefore, it is natural that all three of these distributions exhibit the same kind of behavior in their failure rate $\lambda(t)$ and mathematical expectations $\underline{M}\{\tau_0(t_H)\}$ of the residual lifetime.

It follows from the above reasoning that strengthening leads to a gradual decrease in the rate of wear $\underline{M}\{\xi(t)\}$. However, it would not be correct to assume that, if the rate of wear decreases, strengthening and training necessarily take place. These processes will indeed be observed if, in addition to decrease in the rate of wear, the property of strong mixing of the process of wear is observed.

The situation is analogous with the failure rate $\lambda(t)$. Although the failure rate behaves approximately the same way for the distribution (152) obtained when we have linear wear as for a logarithmic normal distribution, there is still no strengthening in this case. The pretesting observed under ruled sample functions is a consequence of the "burning-in" of items having a high rate of wear.

From what has been said, it is clear that, in analyzing the phenomenon of training, we cannot start with the external form of the

behavior of the rate of wear or the failure rate. Whenever we wish
to look at the reasons for trainability, we need to analyze the
physical picture of the phenomenon. Also, certain bases for making
assumptions regarding the reasons for trainability can be obtained
by experiment on the gradual loading of an object, similar to the
experiments made in the case of cyclic loading of metals [27]. The
nature of this experiment consists in the fact (see Fig. 67) that
the object is trained by a small load S_H over a comparatively extended period of time and then "breaks up" under a relatively high
load S_K. If in the course of a certain time t_H of training it turns
out that the longevity at the level S_K is, on the average, greater
than for nontrained objects and if no items have failed during the
training process, we have a basis for assuming that training of the
objects was attained with strengthening.

The property of trainability and the scheme of stepwise loading
(see Fig. 67) can be used to construct forced tests of longevity
[42, 43].

An analysis of the behavior of the failure rate $\lambda(t)$ is important
in estimating the expediency of preventive maintenance of units. Once
the failure rate $\lambda(t)$ has begun to decrease, it makes no sense to resort to preventive maintenance. Thus, in the case of the failure
rate shown in Fig. 65, preventive maintenance can prove expedient
up to the instant Θ. If, however, the object has already operated
past the instant Θ, preventive maintenance loses meaning.

The fact that the behavior of the failure rate affects in a very
real way an estimate of the expediency of preventive maintenance of
units compels us to use especial care in determining the possibility of using one or the other distribution to describe experimental data regarding the quantities τ_i. Thus, for example, data on
fatigue longevity can, with the same measure of agreement, be
smoothed on the basis of a logarithmic normal distribution and a
Weibull-Gnedenko distribution [40]. However, the behavior of the
failure rates of these distributions is quite different. As a rule,
in smoothing data on fatigue longevity, we obtain the parameter
γ greater than unity. In such a case, the failure rate has a mono-

tonic increase, as we have already pointed out. On the other hand, a logarithmic normal distribution has a nonmonotonic failure rate (see Fig. 65). Conclusions as to the expediency of preventive maintenance of units can be contradictory, depending on the distribution chosen.

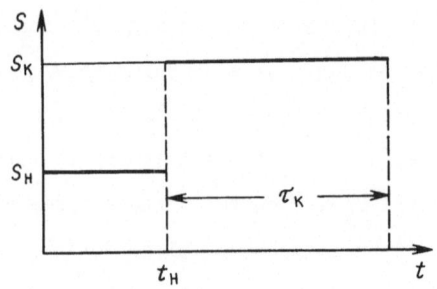

Fig. 67. A scheme of step loading.

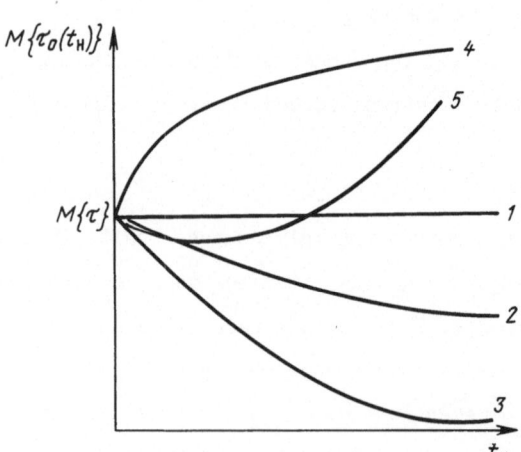

Fig. 68. Qualitative nature of the behavior of $\underline{M}\{\tau_0(t_H)\}$ for different distributions: 1, an exponential distribution; 2, a gamma distribution; 3, a normal distribution or a Weibull-Gnedenko distribution with $\gamma > 1$; 4, a Weibull-Gnedenko distribution with $\gamma < 1$; 5, a logarithmic normal distribution.

To avoid crude errors, it is useful, when we have several hundred data regarding the quantities τ_i, to resort not only to an analysis and smoothing of the polygon of the distribution but also to an analysis of the behavior of the mathematical expectation of the remainder of the longevity $\underline{M}\{\tau_0(t_H)\}$. Fig. 68 shows graphs illustrating the nature of the change in $\underline{M}\{\tau_0(t_H)\}$ for various types of distributions.

Conclusion

The picture of a failure is, as a rule, extremely complicated and our knowledge of its physical nature is more or less relative. However, we cannot always track down all factors influencing a failure. Therefore, our model of the occurrence of a failure is always to some degree approximate. This means that the distribution law of the lifetime τ that we take reflects only certain features of the phenomenon observed. The difficulties that arise in this connection compel us to take into account not only the physics of the phenomenon of failure but also the specific requirement of the problem being solved. For by no means all problems of calculating reliability do we need to have detailed information regarding the physical picture of a failure or the form of the distribution of the lifetime. It would be basically incorrect to assert that, every time one solves problems of the analysis and calculation of reliability, we need to determine as precisely as possible the form of the distribution law of the lifetime. It is sufficient to show that, if the mean lifetime is of practical interest, then, to estimate it from the data $\tau_1, \tau_2, \ldots, \tau_N$, it is not at all necessary to establish the distribution law since an estimate of the mean lifetime will be given by the arithmetic mean $\bar{\tau}$.

However, if we always neglect the necessity of getting as exact a description as possible of the distribution of the quantity τ, we can fall into gross errors and cause a significant amount of damage to the reliability of the unit. To see this, let us use the data of Tab. 4 and show that, if we process these data on the basis of various distribution laws of the lifetime, we shall obtain completely different estimates of the reliability.

Suppose that we are required to find, on the basis of 22 experimental data, the probability of failure in the interval $(0; 0.45 \cdot 10^5)$ of cycles and the time $T_{0.01}$ and the end of which 1% of the units (quantile of level 0.01) have failed. (We have in mind the result of tests of the longevity under a heavy load $\sigma_{max} = 30$ kg/mm^2, Tab. 4). Let us suppose that the picture of a failure has not been studied and we are making a formal smoothing of the data in accordance with several distribution laws. For the normal and logarithmic normal distributions, the parameters were found from formulas (135). For the remaining distributions they were found by the method of discrimination partitions for the values $\Theta_1 = 1 \cdot 10^5; \Theta_2 = 2 \cdot 10^5$. The results of this smoothing are shown in Tab. 11. All the distributions mentioned in it exhibit good agreement with the experimental data. The estimate of agreement was made with the aid of Kolmogorov's criterion [4, p. 126]. The maximum discrepancy between the empirical and theoretical distribution did not exceed 0.12 for the first two distributions of Tab. 11; for the others, it did not exceed 0.17. These deviations are considerably less than the maximum admissible deviations at the 20 % significance level [4,p.409].

Table 11.

The results of smoothing of the data of Table 4 for various distributions.

Form of distribution	Parameters	$\underline{P}\{\tau \leq 0.45 \cdot 10^5\}$	$T_{0.01}$
Logarithmic normal	$c = 5.084$ $\sigma = 0.189$	0.011	$4.4 \cdot 10^4$
Bernstein's distribution with parameter b = 0	$c = 1.21 \cdot 10^5$ $\sqrt{a} = 0.437$	0.00005	$6 \cdot 10^4$
Normal untruncated	$c = 1.33 \cdot 10^5$ $\sigma = 0.6 \cdot 10^5$	0.071	Less than zero
Weibull-Gnedenko	$\gamma = 2.15$ $\beta = 2.61$	0.067	$0.18 \cdot 10^4$

The parameters γ and β are calculated on a time scale with unity corresponding.

It is clear from the table that an estimate of the probability of failure-free operation differs sharply for the different distributions. Fluctuations in the quantities $T_{0.01}$ are not especially great but, in the case in which we are using them as a warranty time, they can have a significant influence on the calculated reliability of a unit.

Here, we note that the method for determining the parameters of the distribution influences the value of the quantities $T_{0.01}$ and $P\{\tau \le 0.45 \cdot 10^5\}$ to a considerably weaker degree than the choice of distribution on the basis of which the experimental data are processed.

Consequently, it would be extremely desirable to use the physical picture of a failure and to make a choice of a finite distribution on the basis of it. We include Tab. 12 for this purpose. Tab. 12 classifies the problems of calculating the reliability and it gives indications of the acceptability of the different distribution laws

Table 12.

Classification of problems of calculating reliability.

Type of problem	Type of distribution
Estimate of the mean lifetime	—
Estimates of the coefficient of use	—
Estimation of the coefficient of variation of the lifetime	—
Calculation of multiple-machine servicing	Exponential
Calculation of the number of a maintenance standby	Exponential
Calculation of the probability of failure of a complicated system after multiple replacement of defective units	Exponential
Calculation of preventive maintenance of units in the case of a monotonically increasing failure rate	Gamma distribution
Calculation of the probability of failure of a large chain system with independent units	Weibull-Gnedenko
Estimation of the probability of failure-free operation when the number of items tested is small	M
Calculation of unloaded standby (cold reserve) in the case of a monotonically increasing failure rate	Gamma distribution
Forced and accelerated tests of reliability	M

for the lifetime. The table was constructed as follows: If it is immaterial which distribution law we use to solve the problem, a dash appears in the column entitled "Type of distribution". If we take some standard type of law independently of the actual distribution law of the time τ, this type is indicated. If it is necessary to have detailed knowledge of the physical picture and to construct a model of the failure, this is indicated by the letter $\underline{M}$.

Of course, this table is not exhaustive, but we can see from it that situations are rather rare in which we need to have a detailed knowledge of the physical picture or a model of the failure. Primarily, these are situations associated with tests of reliability. Thus, if we can use a certain set of standard distributions in the course of planning a system, then, in tests of systems and their parts, we need to look into the physical nature of the failure.

Appendix

Table A1.

Laplace's function $\Phi(x) = \dfrac{1}{\sqrt{2\pi}} \int\limits_{-\infty}^{x} e^{-\frac{u^2}{2}}\, du$.

Hundredth parts of x

x	0	1	2	3	4	5	6	7	8	9
0.0	0.5000	0.5040	0.5080	0.5120	0.5160	0.5199	0.5239	0.5279	0.5319	0.5359
0.1	0.5398	0.5438	0.5478	0.5517	0.5557	0.5596	0.5636	0.5675	0.5714	0.5753
0.2	0.5793	0.5832	0.5871	0.5910	0.5948	0.5987	0.6026	0.6064	0.6103	0.6141
0.3	0.6179	0.6217	0.6255	0.6293	0.6331	0.6368	0.6406	0.6443	0.6480	0.6517
0.4	0.6554	0.6591	0.6628	0.6664	0.6700	0.6736	0.6772	0.6808	0.6844	0.6879
0.5	0.6915	0.6950	0.6985	0.7019	0.7054	0.7088	0.7123	0.7157	0.7190	0.7224
0.6	0.7257	0.7291	0.7324	0.7357	0.7389	0.7422	0.7454	0.7486	0.7517	0.7549
0.7	0.7580	0.7611	0.7642	0.7673	0.7703	0.7734	0.7764	0.7794	0.7823	0.7852
0.8	0.7881	0.7910	0.7939	0.7967	0.7995	0.8023	0.8051	0.8078	0.8106	0.8133
0.9	0.8159	0.8186	0.8212	0.8238	0.8264	0.8289	0.8315	0.8340	0.8365	0.8389
1.0	0.8413	0.8437	0.8461	0.8485	0.8508	0.8531	0.8554	0.8577	0.8599	0.8621
1.1	0.8643	0.8665	0.8686	0.8708	0.8729	0.8749	0.8770	0.8790	0.8810	0.8830
1.2	0.8849	0.8869	0.8888	0.8907	0.8925	0.8944	0.8962	0.8980	0.8997	0.9015
1.3	0.9032	0.9049	0.9066	0.9082	0.9099	0.9115	0.9131	0.9147	0.9162	0.9177

x	0	1	2	3	4	5	6	7	8	9
1.4	0.9192	0.9207	0.9222	0.9236	0.9251	0.9265	0.9279	0.9292	0.9306	0.9319
1.5	0.9332	0.9345	0.9357	0.9370	0.9382	0.9394	0.9406	0.9418	0.9429	0.9441
1.6	0.9452	0.9463	0.9474	0.9484	0.9495	0.9505	0.9515	0.9525	0.9535	0.9545
1.7	0.9554	0.9564	0.9573	0.9582	0.9591	0.9599	0.9608	0.9616	0.9625	0.9633
1.8	0.9641	0.9649	0.9656	0.9664	0.9671	0.9678	0.9686	0.9693	0.9699	0.9706
1.9	0.9713	0.9719	0.9726	0.9732	0.9738	0.9744	0.9750	0.9756	0.9761	0.9767
2.0	0.9772	0.9773	0.9783	0.9788	0.9793	0.9798	0.9803	0.9808	0.9812	0.9817
2.1	0.9821	0.9825	0.9830	0.9834	0.9838	0.9842	0.9846	0.9850	0.9854	0.9857
2.2	0.9861	0.9864	0.9868	0.9871	0.9875	0.9878	0.9881	0.9884	0.9887	0.9890
2.3	0.9893	0.9896	0.9898	0.9901	0.9904	0.9906	0.9909	0.9911	0.9913	0.9916
2.4	0.9918	0.9920	0.9922	0.9925	0.9927	0.9929	0.9931	0.9932	0.9934	0.9936
2.5	0.9938	0.9940	0.9941	0.9943	0.9945	0.9946	0.9948	0.9949	0.9951	0.9952
2.6	0.9953	0.9955	0.9956	0.9957	0.9959	0.9960	0.9961	0.9962	0.9963	0.9964
2.7	0.9965	0.9966	0.9967	0.9968	0.9969	0.9970	0.9971	0.9972	0.9973	0.9974
2.8	0.9974	0.9975	0.9976	0.9977	0.9977	0.9978	0.9979	0.9979	0.9980	0.9981
2.9	0.9981	0.9982	0.9982	0.9983	0.9984	0.9984	0.9985	0.9985	0.9986	0.9986
3.0	0.9987									

For negative values of the argument, $\Phi(-x) = 1 - \Phi(x)$. For example, let x = -0.53. Then $\Phi(-0.53) = 1 - \Phi(0.53) = 1 - 0.7019 = 0.2981$.

Table A2.

The inverse Laplace function ψ(ν).

ν	Ψ(ν)	ν	Ψ(ν)	ν	Ψ(ν)	ν	Ψ(ν)
0.01	-2.3263	0.26	-0.6433	0.51	0.0251	0.76	0.7063
0.02	-2.0537	0.27	-0.6128	0.52	0.0502	0.77	0.7388
0.03	-1.8808	0.28	-0.5828	0.53	0.0753	0.78	0.7722
0.04	-1.7507	0.29	-0.5534	0.54	0.1004	0.79	0.8064
0.05	-1.6449	0.30	-0.5244	0.55	0.1257	0.80	0.8416
0.06	-1.5548	0.31	-0.4958	0.56	0.1510	0.81	0.8779
0.07	-1.4758	0.32	-0.4677	0.57	0.1764	0.82	0.9154
0.08	-1.4051	0.33	-0.4399	0.58	0.2019	0.83	0.9542
0.09	-1.3408	0.34	-0.4125	0.59	0.2275	0.84	0.9945
0.10	-1.2816	0.35	-0.3853	0.60	0.2533	0.85	1.0364
0.11	-1.2265	0.36	-0.3585	0.61	0.2793	0.86	1.0803
0.12	-1.1750	0.37	-0.3319	0.62	0,3055	0.87	1.1264
0.13	-1.1264	0.38	-0.3055	0.63	0.3319	0.88	1.1750
0.14	-1.0803	0.39	-0.2793	0.64	0.3585	0.89	1.2265
0.15	-1.0364	0.40	-0.2533	0.65	0.3853	0.90	1.2816
0.16	-0.9945	0.41	-0.2275	0.66	0.4125	0.91	1.3408
0.17	-0.9542	0.42	-0.2019	0.67	0.4399	0.92	1.4051
0.18	-0.9154	0.43	-0.1764	0.68	0.4677	0.93	1.4758
0.19	-0.8779	0.44	-0.1510	0.69	0.4958	0.94	1.5548
0.20	-0.8416	0.45	-0.1257	0.70	0.5244	0.95	1.6449
0.21	-0.8064	0.46	-0.1004	0.71	0.5534	0.96	1.7507
0.22	-0.7722	0.47	-0.0753	0.72	0.5828	0.97	1.8808
0.23	-0.7388	0.48	-0.0502	0.73	0.6128	0.98	2.0537
0.24	-0.7063	0.49	-0.0251	0.74	0.6433	0.99	2.3263
0.25	-0.6745	0.50	0.0000	0.75	0.6745		

le A3.

The functions $y_1(x) = \ln \dfrac{1}{1-x}$ and $y_2(x) = \ln\ln \dfrac{1}{1-x}$

x	$y_1(x)$	$y_2(x)$	x	$y_1(x)$	$y_2(x)$	x	$y_1(x)$	$y_2(x)$	x	$y_1(x)$	$y_2(x)$
0.01	0.010	-4.600	0.26	0.301	-1.200	0.51	0.713	-0.338	0.76	1.427	0.356
0.02	0.020	-3.902	0.27	0.315	-1.156	0.52	0.734	-0.309	0.77	1.470	0.385
0.03	0.030	-3.491	0.28	0.329	-1.113	0.53	0.755	-0.281	0.78	1.514	0.415
0.04	0.041	-3.199	0.29	0.342	-1.072	0.54	0.777	-0.253	0.79	1.561	0.445
0.05	0.051	-2.970	0.30	0.357	-1.031	0.55	0.799	-0.225	0.80	1.609	0.476
0.06	0.062	-2.783	0.31	0.371	-0.991	0.56	0.821	-0.197	0.81	1.661	0.507
0.07	0.073	-2.623	0.32	0.386	-0.953	0.57	0.844	-0.170	0.82	1.715	0.539
0.08	0.083	-2.484	0.33	0.400	-0.915	0.58	0.868	-0.142	0.83	1.772	0.572
0.09	0.094	-2.361	0.34	0.416	-0.878	0.59	0.892	-0.115	0.84	1.833	0.606
0.10	0.105	-2.250	0.35	0.431	-0.842	0.60	0.916	-0.087	0.85	1.897	0.640
0.11	0.117	-2.150	0.36	0.446	-0.807	0.61	0.942	-0.060	0.86	1.966	0.676
0.12	0.128	-2.057	0.37	0.462	-0.772	0.62	0.968	-0.033	0.87	2.040	0.713
0.13	0.139	-1.971	0.38	0.478	-0.738	0.63	0.994	-0.058	0.88	2.120	0.752
0.14	0.151	-1.892	0.39	0.494	-0.705	0.64	1.022	0.021	0.89	2.207	0.792
0.15	0.163	-1.817	0.40	0.511	-0.672	0.65	1.050	0.049	0.90	2.303	0.834
0.16	0.174	-1.747	0.41	0.528	-0.639	0.66	1.079	0.076	0.91	2.408	0.879
0.17	0.186	-1.680	0.42	0.545	-0.607	0.67	1.109	0.103	0.92	2.526	0.927
0.18	0.198	-1.617	0.43	0.562	-0.576	0.68	1.139	0.131	0.93	2.659	0.978
0.19	0.211	-1.557	0.44	0.580	-0.545	0.69	1.171	0.158	0.94	2.813	1.034
0.20	0.223	-1.500	0.45	0.598	-0.514	0.70	1.204	0.186	0.95	2.996	1.097
0.21	0.236	-1.445	0.46	0.616	-0.484	0.71	1.238	0.213	0.96	3.219	1.169
0.22	0.248	-1.392	0.47	0.635	-0.454	0.72	1.273	0.241	0.97	3.507	1.255
0.23	0.261	-1.342	0.48	0.654	-0.425	0.73	1.309	0.270	0.98	3.912	1.364
0.24	0.274	-1.293	0.49	0.673	-0.395	0.74	1.347	0.298	0.99	4.605	1.527
0.25	0.288	-1.246	0.50	0.693	-0.367	0.75	1.386	0.327			

Bibliography

1. B a z o v s k i y , I.: Nadezhnost', Teoriya i praktika (Reliability, theory and practice), Moscow: MIR, 1965.
2. B e r n s h t e y n , S. N.: Teoriya veroyatnostey (Probability theory),
   OGIZ, 1946.
3. B o l o t i n , V. V.: Statisticheskiye metody v stroitel'noy mek-
   hanike (Statistical methods in construction mechanics),
   Stroyizdat, 1965.
4. B o l ' s h e v , L. N., and S m i r n o v , N. V.: Tablitsy matematicheskoy statistiki (Tables of mathematical statistics), Moscow: Nauka, 1965.
5. B r u y e v i c h , N. G., and S e r g e y e v , V. I.: Nekotoryye
   obshchiye voprosy tochnosti i nadezhnosti ustroystv (Certain
   overall questions of accuracy and reliability of devices) in
   the collection "O tochnosti i nadezhnosti v avtomatizirovannom
   mashinostroyenii", Moscow: Nauka, 1964.
6. B r u y e v i c h , N. G. and G r a b o v e t s k i y , V. P.: Ob osnov-
   nykh napravleniyakh teorii nadezhnosti (On the basic directions
   taken in reliability theory) in the collection "Kiberbetika na
   sluzhbu kommunizmu" (Cybernetics in the service of communism),
   Energiya, 1964.
7. B u s h , R., and M o s t e l l e r , F.: Stochastic Models for Learn-
   ing, New York: Wiley, 1955.
8. G e r t s b a k h , I. B.: O nekotorykh veroyatnostnykh modelyakh
   vozniknoveniya otkazov (On certain probabilistic models of
   occurrence of failures) in the collection "Issledovaniye nadezhnosti apparatury svyazi i radiopriyemnykh ustroystv", Riga:
   LatINTI, 1963.

9. G n e d e n k o , B. V.: Predel'nyye teoremy dlya maksimal'nogo chlena variatsionnogo ryada (Limit theorems for the maximum term of a variational series). Doklady Akad. nauk. USSR, 32, 1941.

10. G n e d e n k o , B. V.: Theory of probability (Translation of "Kurs teorii veroyatnostey"), New York: Chelsea.

11. G n e d e n k o , B. V., B e l y a y e v , Yu. K., and S o l o v ' y e v , A. D.: Mathematical methods in the theory of reliability (Translation of "Matematicheskiye metody v teorii nadezhnosti"), New York: Academic Press (in Press).

12. I b r a g i m o v , I. A.: Nekotoryye predel'nyye teoremy dlya statsionarnykh protsessov (Some limit for stationary processes). Teoriya veroyatnostey i yeye primeneniya, No. 4, 1962.

13. K a g a n , A. M.: Semeystva raspredeleniy i razdelyayushchiye razbiyeniya (Families of distributions and separating partitions). Doklady Akad. nauk USSR, 153, No. 4, 1941.

14. K o r d o n s k i y , Kh. B.: Prilozheniya teorii veroyatnostey v inzhenernom dele (Applications of probability in engineering), Moscow: Fizmatgiz, 1963.

15. K o r d o n s k i y , Kh. B.: Raschet i ispytaniya ustalostnoy dolgovechnosti (Calculation and tests of fatigue longevity), Trudy 4-go matematicheskogo s''yezda (Proceedings of the fourth mathematical congress) Vol. 2, Leningrad, 3-12 June, 1961, Moscow: Nauka 1964.

16. K o r d o n s k i y , Kh. B.: Forsirovannyye ispytaniya nadezhnosti mashin i priborov (Forced tests of reliability of machines and devices). Standartizatsiya, No. 7, 1964.

17. K o r d o n s k i y , Kh. B.: Ustalostnaya dolgovechnost' v svete obshchey teorii iznashivaniya (Fatigue longevity in the light of the general theory of wear) in the collection of addresses of the Soveshchaniya po statisticheskim voprosam prochnosti (Meeting on statistical question of durability), Leningrad, 1964.

18. K o s t e t s k i y , B. I.: Soprotivleniye iznashivaniyu detaley mashin (Resistance of machine parts to wear), Moscow: Mashgiz, 1959.

19. K r a g e l ' s k i y , I. V.: Treniye i iznos (Friction and wearing), Moscow: Mashgiz, 1962.

20. L e o n t ' y e v , L. P., and M a r g u l i s , A. M.: Nadezhnost' i srok sluzhby nekotorykh elektronnykh lamp (Reliability and length of service of certain vacuum tubes), Trudy Instituta elektronika i vychislitel'noy tekhniki AN Lat-oy SSR (Proceedings of the Institute of electronic and computational technology of the Academy of Sciences of the Latvian SSR), No. 5, 1963.

21. L e o n t ' y e v , L. P.: Vvedeniye v teoriyu nadezhnosti radioelektronnoy apparatury (Introduction to the theory of reliability of radioelectronic apparatus), Riga: AN Latv. SSR, 1963.

22. L i n n i k , Yu. V.: Method of least squares and principles of the theory of observations (Translation of "Metod naimen'shikh kvadratov i osnovy teorii obrabotki nablyudeniy"), New York: Pergamon, 1961.

23. Semiconductor Reliability. Edited by W. H. V o n A l v e n . New Jersey: Engineering Publishers Elizabeth 1962.

24. P a r o l ' , N. V.: Nadezhnost' priyemno-usilitel'nykh lamp (Reliability of receiving-amplifying tubes), Moscow: Sovetskoye radio, 1964.

25. P o l o v k o , A. M.: Reliability theory (Translation of "Osnovy teorii nadezhnosti"), New York: Academic Press.

26. P u g a c h e v , V. S.: Theory of random functions and its application to Control Problems (Translation of "Teoriya sluchaynykh funktsiy i yeye primeneniye k zadacham avtomaticheskogo upravleniya"), Reading Massachusetts: Addison-Wesley, 1965.

27. S e r e n s e n , S. V., K a g a y e v , V. P., and S h n e y d e r o v i c h R. M.: Nesushchaya sposobnost' i raschety detaley mashin na prochnost' (Bearing capacity and calculations of parts of machines for durability), Moscow: Mashgiz, 1963.

28. D u n i n - B a r k o v s k i y , I. V., and S m i r n o v , N. V.: Kurs teorii veroyatnostey i matematicheskoy statistiki (Course of probability theory and mathematical statistics), Moscow: Nauka, 1965.

29. T o m a s h o v , N. D.: Theory of corrosion and protection of met-

als (Translation of "Teoriya korrozii i zashchity metallov"),
New York: McMillan, 1965.

30. F e l l e r, W.: Introduction to probability theory and its applications, New York: Wiley, 1957-1966.

31. H a l d, A.: Statistical theory with engineering applications,
New York: Wiley, 1952.

32. K h i n c h i n, A. Ya.: Matematicheskiye metody teorii massovogo
metody teorii massovogo obsluzhivaniya (Mathematical methods in
the theory of mass servicing), Trudy Matemat. in-ta AN (Proceedings of the mathematical institute of the Academy of Sciences), USSR, 1955.

33. K h r u s h c h o v, M. M.: "Klassifikatsiya usloviy i vidov iznashivaniya detaley mashin" (Classification of conditions and forms
of wear of parts of machines) in the collection "Treniye i iznos
v mashinakh" (Friction and wear in machines). Akad. nauk. USSR,
1953.

34. S h i s h o n o k, N. A., R e p k i n, V. F., and B a r v i n s k i y,
L. L.: Osnovy teorii nadezhnosti (Fundamentals of reliability
theory), Moscow: Sovetskoye radio, 1964.

35. S h o r, Ya. B.: Statisticheskiye metody analiza i kontrolya
kachestva i nadezhnosti (Statistical methods of analysis and
control of quality and reliability), Moscow: Sovetskoye radio,
1962.

36. A i t c h i s o n, J., and B r o w n, J. A.: Lognormal distribution,
Cambridge University Press, 1957.

37. B e r r e t t o n i, J. N.: Practical applications of the Weibull
distribution, Ind. Qual. Control, August, 1964.

38. C o x, D. R.: Renewal theory. London: Methuen, and New York:
Wiley, 1962.

39. W e i b u l l, W.: A statistical theory of strength of materials.
Proc. Roy. Swed. Inst. Eng. Research, 15, 1939.

40. W e i b u l l, W.: Scatter of fatigue life and fatigue strength
in aircraft structural materials and parts. Fatigue in aircraft structures, No. 4, 1956.

41. G e r t s b a k h, I. B.: Ob otsenke parametrov raspredeleniya s

eksponentsial'nym mnozhitelem (On estimating the parameters of a distribution with an exponential factor). Teoriya veroyatnostey i yeye primeneniya, 12, No. 1, 1967.

42. K o r d o n s k i y, Kh. B., and F r e s i n, B. S.: Forsirovannyye na ustalostnuyu dolgovechnost' metodom dolamyvaniya (Forced tests on fatigue longevity by the method of testing to destruction), Zavodskaya laboratoriya, No. 3, 1967.

43. K o r d o n s k i y, Kh. B., and F r e s i n, B. S.: Forsirovannyye ispytaniya nadezhnosti mashin (Forced tests of reliability of machines) in the collection "Nadezhnost' i dolgovechnost' mashin i priborov" (Reliability and Longevity of Machines and devices), NiiMash, No. 7, 1966.

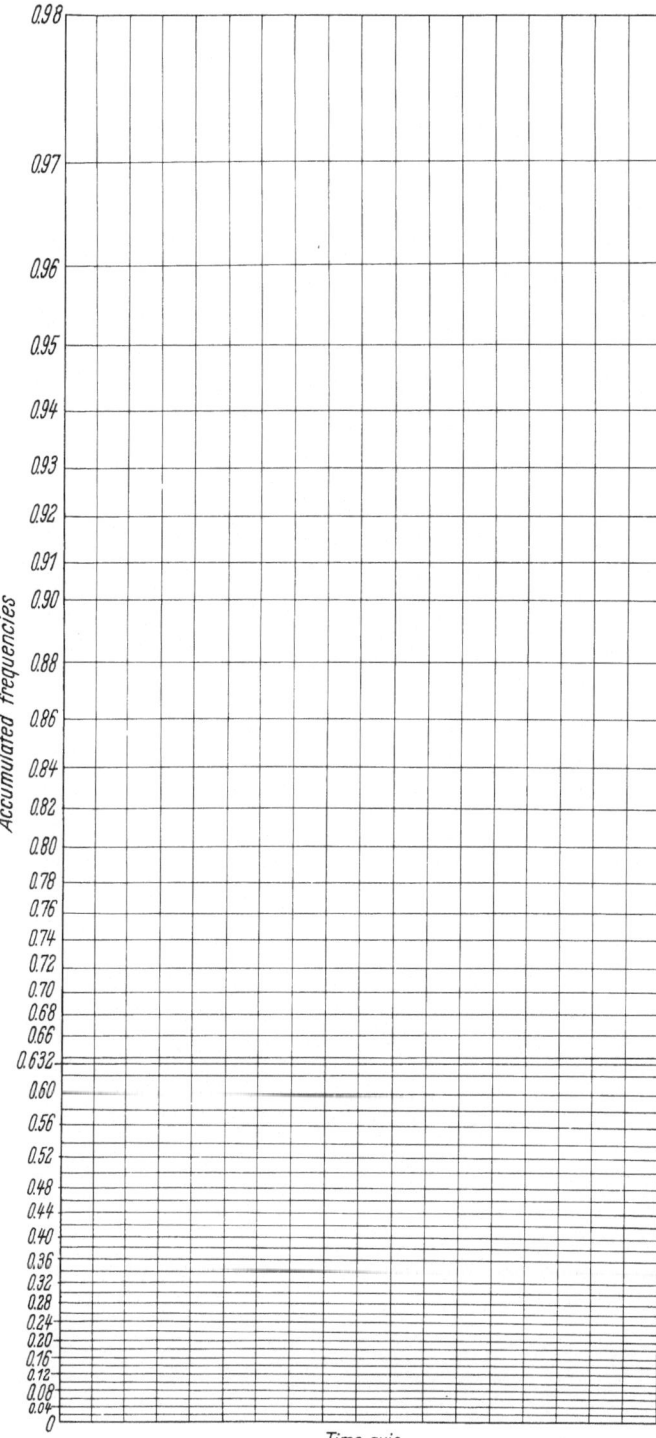

Probability paper for an exponential distribution (for directions see p. 23).

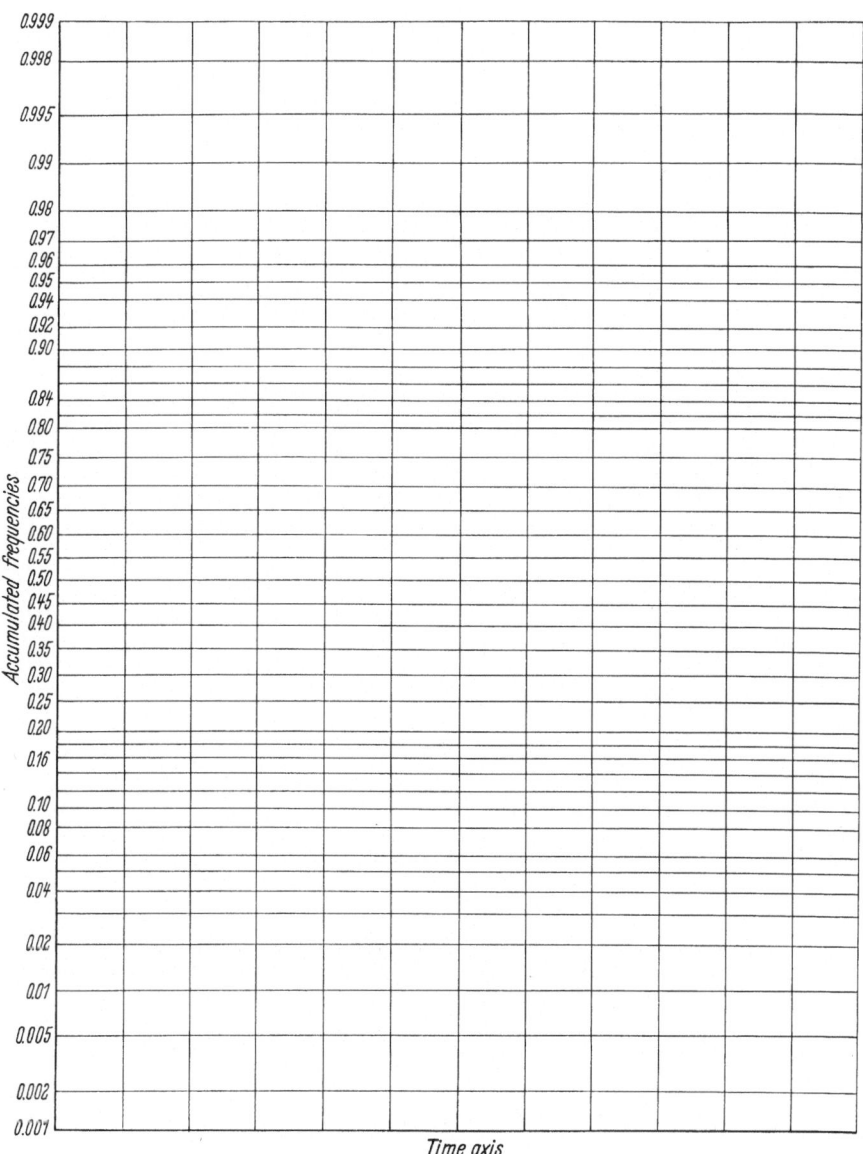

Probability paper for an normal distribution (for directions see p. 23, 64).

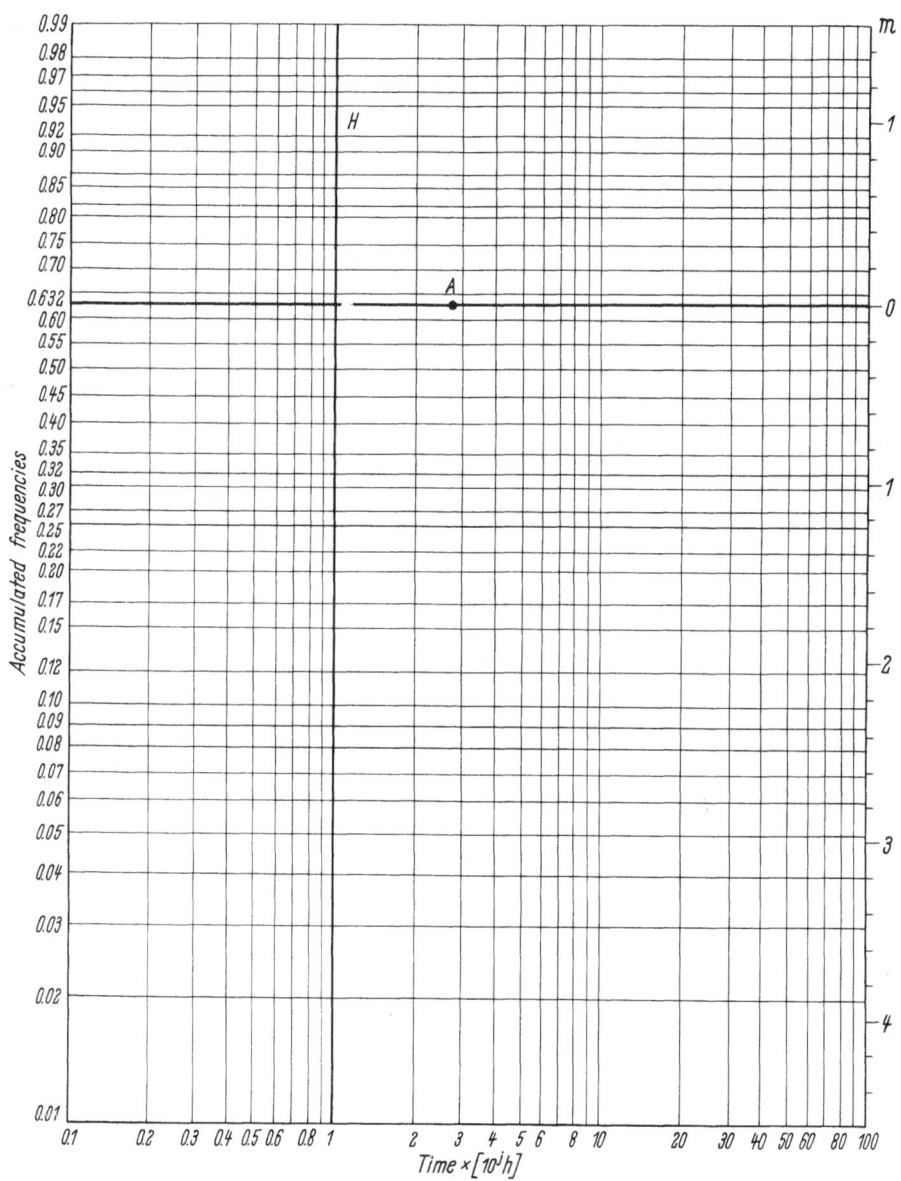

Probability paper for a Weibull-Gnedenko distribution (for directions see p. 23, 125-127).